ERGEBNISSE DER MATHEMATIK UND IHRER GRENZGEBIETE

UNTER MITWIRKUNG DER SCHRIFTLEITUNG DES „ZENTRALBLATT FÜR MATHEMATIK“

HERAUSGEGEBEN VON

L. V. AHLFORS · R. BAER · R. COURANT · J. L. DOOB · S. EILENBERG
P. R. HALMOS · T. NAKAYAMA · H. RADEMACHER
F. K. SCHMIDT · B. SEGRE · E. SPERNER

NEUE FOLGE · HEFT 10

REIHE:

GRUPPENTHEORIE

BESORGT

VON

R. BAER

SPRINGER-VERLAG
BERLIN · GÖTTINGEN · HEIDELBERG
1956

STRUCTURE OF A GROUP AND THE STRUCTURE OF ITS LATTICE OF SUBGROUPS

BY

MICHIO SUZUKI

SPRINGER-VERLAG
BERLIN · GÖTTINGEN · HEIDELBERG
1956

ISBN 978-3-642-52760-9 ISBN 978-3-642-52758-6 (eBook)
DOI 10.1007/978-3-642-52758-6

Preface.

The central theme of this monograph is the relation between the structure of a group and the structure of its lattice of subgroups. Since the first papers on this topic have appeared, notably those of BAER and ORE, a large body of literature has grown up around this theory, and it is our aim to give a picture of the present state of this theory. To obtain a systematic treatment of the subject quite a few unpublished results of the author had to be included. On the other hand, it is natural that we could not reproduce every detail and had to treat some parts somewhat sketchily.

We have tried to make this report as self-contained as possible. Accordingly we have given some proofs in considerable detail, though of course it is in the nature of such a report that many proofs have to be omitted or can only be given in outline. Similarly references to the concepts and theorems used are almost exclusively references to standard works like BIRKHOFF [*1*] and ZASSENHAUS [*1*].

The author would like to express his sincere gratitude to Professors REINHOLD BAER and DONALD G. HIGMAN for their kindness in giving him many valuable suggestions. His thanks are also due to Dr. NOBORU ITÔ who, during stimulating conversations, contributed many useful ideas.

Urbana, May, 1956.

M. Suzuki.

Contents.

Notations.

$G, G', \ldots$:	groups, finite or not.
$H, K, U, V, \ldots$:	subgroups of a group to be considered.
e:	the unit subgroup, consisting of the identity element only.
$a, b, c, \ldots$:	elements in groups.
1:	the identity element of a group.
$\{a, b, \ldots\}$:	the subgroup generated by $a, b, \ldots$.
Λ:	an index set, finite or not.
λ:	the general element of Λ.
G^n:	the subgroup of G generated by all n-th powers of elements.
$C(G)$:	the commutator subgroup of a group G.
$L(G)$:	the subgroup lattice of a group G.
$L\{a, b, \ldots\}$:	$= L(H)$, where $H = \{a, b, \ldots\}$.
$N_p(G)$:	the intersection of all normal subgroups of a finite group G whose factor groups are p-groups.
$R(G)$:	the radical, the maximal solvable normal subgroup of a finite group G.
$S(G)$:	the intersection of all normal subgroups of a finite group G with solvable factor groups.
$Z(G)$:	the center of G.
$\Phi(G) = \Phi$:	the Φ-subgroup, the intersection of all maximal subgroups of G.
$(U:V)$:	the index of V in U.
$U \cap V$:	the intersection of two subgroups U and V.
$U \cup V$:	the subgroup generated by U and V.
$U\,V$:	$= U \cup V$, when U and V are permutable.
$U \times V$:	the direct product of U and V.
$\Pi_\lambda U_\lambda$:	the direct product of subgroups $U_\lambda (\lambda \in \Lambda)$.
$U \supset V$:	V is a proper part of U.
$U \supseteq V$:	V is a subgroup of U.
$\varphi, \theta \cdots$:	mappings defined on subgroup lattices.
$\sigma, \tau \cdots$:	mappings defined on group elements.
$\varphi(H)$:	the image of a subgroup H by φ.
$\varphi\{a, b, \ldots\}$:	$= \varphi(H)$, where $H = \{a, b, \ldots\}$.
I:	maximal elements in lattices.
O:	minimal elements in lattices.

Introduction.

We are interested in the lattice formed by the totality of subgroups of a group. Defining the meet and the join of subgroups of a group G in the natural way, one sees easily that the totality of subgroups of a group G satisfies all axioms of a lattice. We shall call this lattice the subgroup lattice of G and denote it by $L(G)$.

One may ask oneself several questions. For instance, (A) what can we say about the structure of groups, if we impose some lattice theoretical restrictions on their subgroup lattices, (B) to what extent is the group determined by its subgroup lattice, (C) are there any characteristic properties of subgroup lattices, etc.?

It is the purpose of this report to collect the results so far obtained in answer to these and related questions concerning subgroup lattices.

In studying various algebraic systems emphasis has been put on the structure of their subsystems rather than on the behavior of individual elements in the systems, and the lattice of subgroups (or admissible subgroups with respect to some operator domain) has drawn the particular attention of mathematicians since the birth of group-theory. It was DEDEKIND who considered the system of ideals in a ring of algebraic integers for the first time from the lattice theoretical point of view, and he discovered and used the modular identity, sometimes called the DEDEKIND law, in his calculation of ideals. But the real history of the theory of subgroup lattices began in 1928, when ADA ROTTLÄNDER considered in her paper [*1*] the totality of subgroups of a finite group and the mappings between subgroup lattices in solving a question arising from field extensions.

The GALOIS theory of field extensions establishes a one-to-one correspondence between the lattice of intermediate fields of an extension and the subgroup lattice of its GALOIS group. Suppose that two normal extensions K_1/k_1 and K_2/k_2 of finite degree are given. If the GALOIS groups of these field extensions are isomorphic, then this isomorphism induces a one-to-one mapping f defined on the totality of intermediate fields of K_1/k_1 onto the totality of intermediate fields of K_2/k_2 such that

(1) $K_1 \supseteq L_1 \supset L_2 \supseteq k_1$ if and only if $K_2 \supseteq f(L_1) \supset f(L_2) \supseteq k_2$,

(2) $[L_1:L_2] = [f(L_1):f(L_2)]$, and

(3) two intermediate fields L_1 and L_2 of K_1/k_2 are conjugate with respect to the third intermediate field L_3, if and only if $f(L_1)$ and $f(L_2)$ are conjugate with respect to $f(L_3)$.

Conversely such a mapping f induces an isomorphism φ of the subgroup lattice of the GALOIS group G_1 of K_1/k_1 onto that of the GALOIS group G_2 of K_2/k_2 with the two additional properties (a) and (b):

(a) the index $(U:V)$ of two subgroups U and V of G_1 is equal to the index $(\varphi(U):\varphi(V))$, and

(b) two subgroups U and V of a subgroup W of G_1 are conjugate in W if and only if $\varphi(U)$ and $\varphi(V)$ are conjugate in $\varphi(W)$.

Now the question arises whether two groups are isomorphic if there exists such an isomorphism between their subgroup lattices. ROTTLÄNDER solved this question affirmatively in the case that at least one of the groups involved is finite and abelian, and negatively in the non-abelian case by giving an example of non-isomorphic finite groups with the same situation of subgroups.

Since then, many investigations by BAER, IWASAWA, ORE, SADOVSKY and others have appeared, which made the relation between the structure of a group and the structure of its subgroup lattice clearer, and many classes of groups, for instance the class of locally free groups, the class of abelian groups which are „not too small", or the class of finite solvable groups, have been characterized by their subgroup lattices. Thus the structure of the subgroup lattice reflects very strongly the structure of the group.

The consideration of subgroup lattices will be an interesting one for its own sake as well as for its applications. Sometimes the lattice theoretical considerations will simplify the nature of problems on pure group theory. As an example we may recall the determination by IWASAWA [*1*] of the structure of quasi-Hamiltonian groups; these are groups in which every pair of subgroups is permutable. The subgroup lattice of such groups is necessarily modular and this fact is used essentially in the proofs. Furthermore, very important classes of finite groups, for instance the direct product of two isomorphic simple groups, are characterized by their subgroup lattices, so that we might have a possibility to apply lattice theoretical considerations to solve the classification problem of finite simple groups. No consideration along this line has been made, and further investigations would be quite desirable.

Chapter I.

Groups with a special kind of subgroup lattice.

1. The distributive law in subgroup lattices.

A pair (A, B) of subgroups A and B of a group G is said to be a *distributive pair*, if the distributive law

$$C \cap (A \cup B) = (C \cap A) \cup (C \cap B)$$

is satisfied by every subgroup C of G (ORE [2]).

Theorem 1. *A pair of subgroups A and B forms a distributive pair if and only if for every element c of $A \cup B$, not in A nor in B, its relative orders with respect to A and B are finite and relatively prime to each other.*

Here the *relative order* of an element c with respect to a subgroup A is the smallest positive integer $n = n_A$ such that $c^n \in A$.

Proof. Suppose that a pair (A, B) is distributive. If an element c of $A \cup B$ is contained in neither A nor B, then we have

$$C = \{c\} = (C \cap A) \cup (C \cap B),$$

where $C \cap A \neq e$ and $C \cap B \neq e$. Hence both $C \cap A$ and $C \cap B$ have finite index in C and the relative orders n_A and n_B of c are finite. Now $c^{n_A} = a$ generates $C \cap A$ and similarly $c^{n_B} = b$ generates $B \cap C$. Since $\{c\} = \{a\} \cup \{b\}$, we have $c = a^x b^y$, or

$$n_A x + n_B y \equiv 1 \pmod{\text{order of } c}.$$

It follows now easily that n_A and n_B are relatively prime.

Conversely assume that the conditions of this theorem are satisfied. Then every cyclic subgroup C of $A \cup B$ is a join of $C \cap A$ and $C \cap B$. The same is true for arbitrary subgroups of $A \cup B$, in particular we have

$$\begin{aligned} C \cap (A \cup B) &= (C \cap (A \cup B) \cap A) \cup (C \cap (A \cup B) \cap B) \\ &= (C \cap A) \cup (C \cap B) \end{aligned}$$

for every subgroup C of G.

A group is called a *D-group* if its lattice of subgroups is distributive. The first main theorem of our discussion is the following theorem of O. ORE [2] (p. 267, theorem 4).

Theorem 2. *A group G is a D-group if and only if every finite set of elements in G generates a cyclic subgroup.*

Proof. Suppose that a group G is a D-group. Consider a subgroup H generated by two elements a and b. Suppose $b\,a\,b^{-1} \notin \{a\}$. Then, by theorem 1, there would exist integers m and n such that

$$b\,a^m\,b^{-1} \in \{a\},\ b\,a^n\,b^{-1} \in \{b\} \text{ and } (m, n) = 1.$$

Since m and n are relatively prime, there exist integers x and y such that $a = a^{mx}\,a^{ny}$. Hence we would have

$$b\,a\,b^{-1} = b\,a^{mx}\,b^{-1}\,b\,a^{ny}\,b^{-1} = b\,a^{mx}\,b^{-1}\,a^{ny} \in \{a\},$$

which contradicts our assumption. Hence both $\{a\}$ and $\{b\}$ are normal in H. Hence the commutator $d = a\,b\,a^{-1}\,b^{-1}$ is contained in $\{a\} \cap \{b\}$, and so commutes with a and b. The relative order of a with respect to $\{b\}$ is equal to that with respect to $\{a\} \cap \{b\}$. Since therefore $\{a\} \cap \{b\} = \{a^k\} = \{b^l\}$ with $(k, l) = 1$, we have $d^k = a^k\,b\,a^{-k}\,b^{-1} = 1$ and $d^l = a\,b^l\,a^{-1}\,b^{-l} = 1$. Since $(k, l) = 1$, we conclude $d = 1$, and hence H is cyclic.

Suppose conversely that every finite set of elements generates a cyclic subgroup. Then G is abelian. Take two subgroups A and B. Every element c of $A \cup B$ is a product $a\,b$ of elements $a \in A$ and $b \in B$. Since $\{a, b\}$ is by assumption cyclic, the relative orders of c with respect to A and B are relatively prime. Hence by theorem 1, G is a D-group. This proves our theorem.

The case in which two disjoint subgroups A and B form a distributive pair will be important. We have the following theorem of Jones [*1*], (p. 547, theorem 2.4).

Theorem 3. *Let (A, B) be a distributive pair such that $A \cap B = e$. Then $A \cup B$ is a direct product of A and B, and the orders of elements in A and in B are finite and relatively prime.*

Proof. Suppose that A and B form a distributive pair such that $A \cap B = e$. Then every subgroup U of $A \cup B$ is the union of $U \cap A$ and $U \cap B$. Hence the lattice $L(A \cup B)$ of subgroups is the direct product of $L(A)$ and $L(B)$. Take elements $a \in A$ and $b \in B$. Then, by theorem 2, $L\{a\}$ and $L\{b\}$ are both distributive, and so is the direct product $L\{a\} \times L\{b\}$. Since $L\{a, b\} = L\{a\} \times L\{b\}$, the subgroup $H = \{a, b\}$ generated by a and b is cyclic by theorem 2. The generator of H has finite relative orders with respect to both A and B. Since $A \cap B = e$, it is actually of finite order. Hence the order of a and b are finite and relatively prime to each other.

The converse of the above theorem is also true (Iwasawa [*1*], p. 180, Hilfssatz 14). In fact we prove a more general theorem (Suzuki [*1*], p. 346, Lemma 3).

Theorem 4. *The lattice $L(G)$ of subgroups of the group G is a direct product of several lattices $L_\lambda (\lambda \in \Lambda): L(G) = \Pi_{\lambda \in \Lambda} L_\lambda$ if and only if G is a direct product of groups G_λ such that $L(G_\lambda) \cong L_\lambda (\lambda \in \Lambda)$ and the order of any element of G_λ is finite and relatively prime to the order of any element of $G_\mu (\lambda \neq \mu)$.*

Proof. Suppose first $L(G)$ to be decomposable. Let φ be an isomorphism of $L = \Pi_{\lambda \in \Lambda} L_\lambda$ onto $L(G)$. Every element of L is a vector $(\ldots, a_\lambda, \ldots)$ with one component a_λ from each L_λ. Since $L(G)$ has both a greatest and a least element, each L_λ has a greatest element I_λ and a least one O_λ. Consider an element $\bar{I}_\lambda = (, \ldots, O_\mu, \ldots, I_\lambda, \ldots)$ of L with O_μ as its μ-component for all $\mu \neq \lambda$. Put $G_\lambda = \varphi(\bar{I}_\lambda)$. Then G_λ and $G_\mu (\lambda \neq \mu)$ form a distributive pair such that $G_\lambda \cap G_\mu = e$. By theorem 3, we conclude the necessity of the conditions of this theorem.

Assume conversely that the conditions are satisfied. Then each subgroup H of G is a direct product of its subgroups $H_\lambda = H \cap G_\lambda$. Hence if $H = \Pi H_\lambda$ and $K = \Pi K_\lambda$, we have $H \cup K = \Pi (H_\lambda \cup K_\lambda)$ and $H \cap K = \Pi (H_\lambda \cap K_\lambda)$. This shows that $L(G) \cong \Pi L(G_\lambda)$.

A group whose lattice of subgroups is decomposable into a non-trivial direct product is sometimes said to be *L-decomposable.* An L-decomposable group is a torsion group and directly decomposable (theorem 4).

The Φ-subgroup $\Phi(G)$ of a finite group G is the intersection of all maximal subgroups of G. $\Phi(G)$ is a nilpotent group and $G = \Phi H$ implies $H = G$ for any subgroup H of G (Zassenhaus [*1*], pp. 44—45 and 115). Applying theorems of Sylow and of Schur, we obtain the following proposition.

Proposition 1.1. *A finite group G is L-decomposable if and only if G/Φ is L-decomposable* (Suzuki [*1*], p. 348, Lemma 5).

2. Modular identity in subgroup lattices.

Two subgroups U and V of a group G are said to be *permutable* if $U \cup V = UV = VU$. If U is self-conjugate, then U is permutable with all subgroups V of G.

Theorem 5. *If two subgroups U and V are permutable, then the modular identity*

$$(U \cup V) \cap W = U \cup (V \cap W) \text{ for } W \supseteq U$$

is satisfied.

Proof. Clearly $(U \cup V) \cap W \supseteq U \cup (V \cap W)$. If w is an element of $(U \cup V) \cap W$, then $w \in W$ and $w \in U \cup V$. Hence $w = uv$ $(u \in U, v \in V)$.

By assumption $W \supseteq U$ so that $u^{-1} w = v \in W$. Hence $v \in V \cap W$ and $w = u v \in U \cup (V \cap W)$. This shows that $(U \cup V) \cap W = U \cup (V \cap W)$.

Hence if subgroups U and V are permutable, the interval $U \cup V/U$ is mapped isomorphically into the interval $V/U \cap V$ by the mapping $W \to W \cap V$. The lattice formed by all *normal* subgroups of G is, by theorem 5, a modular lattice.

Two normal subgroups M and N are called ***directly similar*** in H if there exists another normal subgroup L such that $H = M \cup L = N \cup L$ and $M \cap L = N \cap L = e$. In this case M and N are isomorphic to H/L. Moreover every element m of M is a product $n x$ of elements n of N and x of L. Now $m \to n$ is an isomorphism of M onto N, and x is in the center of $M \cup N$. Consequently we have shown the following

Proposition 1.2. *If two normal subgroups M and N are directly similar, they are centrally isomorphic in $M \cup N$.*

Theorem 6. (ORE [*1*], p. 173, theorem 4). *Let A, B and C be normal subgroups of G. Then the factor group*

$$\big((A \cup B) \cap (B \cup C) \cap (C \cup A)\big) / \big((B \cap C) \cup (C \cap A) \cup (A \cap B)\big)$$

is abelian.

Proof. Let $T(X) = (X \cap (Y \cup Z)) \cup (Y \cap Z)$, where (X, Y, Z) is a permutation of (A, B, C). Then

$$\begin{aligned} T(A) \cup T(B) &= \big(A \cap (B \cup C)\big) \cup \big(B \cap (C \cup A)\big) \cup (B \cap C) \cup (C \cap A) \\ &= (A \cup B) \cap (B \cup C) \cap (C \cup A), \end{aligned}$$

and

$$\begin{aligned} T(A) \cap T(B) &= \big(A \cup (B \cap C)\big) \cap (B \cup C) \cap \big(B \cup (C \cap A)\big) \cap (C \cup A) \\ &= (A \cap B) \cup (B \cap C) \cup (C \cap A). \end{aligned}$$

Similarly

$$T(A) \cup T(B) = T(B) \cup T(C) = T(C) \cup T(A) = U$$

and

$$T(A) \cap T(B) = T(B) \cap T(C) = T(C) \cap T(A) = V.$$

Hence

$$U/V \cong (T(A)/V) \times (T(B)/V).$$

Since $T(A)/V$ and $T(B)/V$ are both in the center of U/V, U/V is abèlian. Theorem 6 is thus proved.

A lattice L is distributive, if and only if L is modular and

$$(a \cup b) \cap (b \cup c) \cap (c \cup a) = (a \cap b) \cup (b \cap c) \cup (c \cap a) \text{ for } a, b, c \in L.$$

Hence if G contains only perfect normal subgroups, then the lattice formed by all normal subgroups is distributive. Groups with a distributive lattice of normal subgroups were studied by KUNTZMAN [*1*]. Among other things he proved ***the existence of finite groups with a given distributive lattice as its normal subgroup lattice.***

The modular identity is closely connected with the notion of permutability of subgroups. This connection was first considered by ORE [*1* and *3*] and subsequently by IWASAWA [*1*].

The subgroups U and V of a finite group G are permutable, if and only if

$$(U : U \cap V) = (U \cup V : V).$$

In particular if the index of U in G is relatively prime to that of V, then U and V are permutable and $G = U \cup V$ (ORE [*3*], p. 436, theorem 8).

A subgroup N of a group G is called *quasi-normal*, if N is permutable with all subgroups of G. The union of an arbitrary number of quasi-normal subgroups is quasi-normal. If N is quasi-normal, then all subgroups conjugate to N are also quasi-normal.

Proposition 1.3. *If N is a maximal quasi-normal subgroup of G, then N is a normal subgroup of G* (ORE [*3*], p. 438, theorem 16).

Proof. Suppose that N is not normal. Then there is a subgroup $M \neq N$ conjugate to N in G. M is also quasi-normal, and so is $M \cup N$. Since N is maximal, we have $M \cup N = G$. But this is impossible, since $m\,n\,N\,n^{-1}\,m^{-1} = M$ would imply $N = n\,N\,n^{-1} = m^{-1}\,M\,m = M$. Hence N is a normal subgroup of G.

A group is called *quasi-Hamiltonian* (or *quasi-abelian*) if all subgroups are quasi-normal. Abelian or Hamiltonian groups are quasi-Hamiltonian. Such groups were studied by IWASAWA [*1*] and ZAPPA [*2, 3, 4, 5*]. By theorem 5 of p. 5, a quasi-Hamiltonian group has a modular lattice of subgroups, and it is nilpotent by (1.3), if the order is finite. IWASAWA [*1*] proved the converse statement ([*1*], p. 184, Satz 7).

Theorem 7. *A finite group is quasi-Hamiltonian, if and only if it is nilpotent and the subgroup lattices of its* SYLOW *subgroups are modular.*

Proof. We have only to show that all p-subgroups U and V of G are permutable. By the modular identity, the interval $U \cup V/V$ has the same dimension as $U/U \cap V$. Since $U \cup V$ is a p-group, this implies that $(U \cup V : V) = (U : U \cap V)$. Hence U and V are permutable. This proves the sufficiency of the conditions.

Two subgroups U and V form a *modular pair*, if the modular identities

$$(U \cup V) \cap W = U \cup (V \cap W) \text{ for } W \supseteq U$$

and

$$(U \cup V) \cap W = V \cup (U \cap W) \text{ for } W \supseteq V$$

are satisfied. These relations do not, in general, guarantee the permutability of U and V. There is, however, a case in which the permutability is a consequence of the modular identities (IWASAWA [1], p. 186, Satz 10).

Theorem 8. *Let G be a finite group. Every modular pair of subgroups is permutable, if and only if G is nilpotent.*

Proof. Assume first that two subgroups U and V are permutable whenever the modular identities are satisfied. Take a p-SYLOW subgroup S of G and consider a maximal subgroup M of G containing S. If M is not normal, a conjugate subgroup L of M is also maximal. We can easily verify the modular identities for $U = M$ and $V = L$. Hence by assumption M and L are permutable, and that this is impossible, is seen as in the proof of (1.3). Hence M is normal, and this implies that G is nilpotent.

Conversely assume that G is nilpotent. G is a direct product of SYLOW subgroups $S_1, \ldots, S_n$. Any subgroup U of G is, therefore, a direct product of its SYLOW subgroups $U_i = U \cap S_i$. Suppose two subgroups $U = \Pi\, U_i$ and $V = \Pi\, V_i$ satisfy the modular identities. Denote the orders of $U, V, U \cup V, U \cap V, U_i, V_i, U_i \cup V_i$ and $U_i \cap V_i$ by $u, v, m, d, u_i, v_i, m_i$ and d_i respectively. By $U \cup V = \Pi\,(U_i \cup V_i)$ and $U \cap V = \Pi\,(U_i \cap V_i)$, it follows that $\Pi\, m_i = m$ and $\Pi\, d_i = d$, and $m\,d$ is divisible by $u\,v$. On the other hand, by the modular identity the interval $U \cup V/U$ of $L(G)$ is mapped isomorphically into $V/U \cap V$. Hence the length of a principal chain connecting $U \cup V$ and U is not larger than that of a principal chain between V and $U \cap V$. Since G is nilpotent, this implies that the number of prime factors in $(U \cup V : U) = m/u$ is not larger than the number of prime factors in $(V : U \cap V) = v/d$. Since $m\,d$ is divisible by $u\,v$, we must have $m\,d = u\,v$, or

$$(U \cup V : U) = (V : U \cap V).$$

Hence U and V are permutable.

A group G is called *permutably decomposable*, if G is a union of two permutable subgroups, i. e. $G = U \cup V = U\,V = V\,U$. The structure of such finite groups has been studied by ORE [3] and more recently by P. HALL, HUPPERT, ITÔ, RÉDEI, SZÉP, WIELANDT, ZAPPA and others.

3. The Jordan-Dedekind chain condition and lower semi-modularity.

A group G is called a *J-group*, if its subgroup lattice $L(G)$ satisfies the JORDAN-DEDEKIND chain condition:

If U and V are subgroups of G and $U \supseteq V$, there exists a series

$$U = H_0 \supset H_1 \supset \cdots \supset H_r = V$$

such that each H_i is a maximal subgroup of H_{i-1} for $i = 1, 2, \ldots, r$, and every such series has the same finite length r.

It is not yet known whether a J-group is always finite or not. We deal in this section only with finite J-groups. The following theorem is due to IWASAWA [*1*] (p. 173, Satz 1).

Theorem 9. *A finite group G is a J-group, if and only if it is supersolvable: i. e. if and only if there exists a principal series whose factors are of order a prime.*

Proof. Assume that G is a J-group. We want to show the existence of a principal series

$$G = N_0 \supset N_1 \supset \cdots \supset N_r = e$$

such that the index $(N_{i-1}:N_i)$ is a prime number p_i and $p_1 \leq p_2 \leq \cdots \leq p_r$. Using induction on the order g of G, we may assume the assertion to be true for groups of smaller order. Let q be the smallest prime factor of g. Suppose G is *q-normal* (ZASSENHAUS [*1*], p. 133). By a theorem of GRÜN, the q-commutator factor group $G/G'(q)$ is isomorphic to $N_z/N_z'(q)$ where N_z is the normalizer of the center of a q-SYLOW subgroup of G. If $N_z \neq G$, then by inductive hypothesis $G \neq G'(q)$, as $N_z \neq N_z'(q)$. If G is not q-normal, by a theorem of BURNSIDE (ZASSENHAUS [*1*], p. 56), there is a q-subgroup H of G such that its normalizer N in G contains an element whose order is relatively prime to q and which induces a non-trivial automorphism in H. By hypothesis of induction N must coincide with G. Hence in any case G contains a proper normal subgroup, and so G must be solvable. Hence the length of a principal chain connecting G and e is equal to the number of prime factors of g. Take the greatest prime factor p of g, and consider a p-SYLOW subgroup S. We refine the chain $G \supseteq S \supset e$ into a maximal one which has the form

$$e = U_0 \subset U_1 \subset \cdots \subset U_k = S \subset U_{k+1} \subset \cdots \subset U_r = G.$$

Since G is a J-group, each index $(U_{i+1}:U_i) = q_i$ $(i = k, k+1, \ldots, r-1)$ is a prime number smaller than p. Applying a theorem of SYLOW we see that S is a normal subgroup of G. Now by HALL's theorem G contains a subgroup H such that $G = S\,H$ and $S \cap H = e$. Consider the center Z of S. Since S is normal, Z is also normal in G. We take a subgroup V of $Z\,H$ such that V covers H. $V \cap Z$ is a normal subgroup of V, and hence it is normal in G. Now we compare the lengths of the two longest chains

$$V \supset H \supset \cdots \supset e$$

and

$$V \supset \cdots \supset V \cap Z \supset \cdots \supset e.$$

Since $H \cong V/V \cap Z$, $V \cap Z$ contains no proper subgroup. Hence G contains a normal subgroup of order p and this completes our proof.

Conversely, suppose G to be super-solvable. Then every subgroup of G is also super-solvable, and every maximal subgroup of any subgroup of G has a prime index. Hence the length of any longest chain of $L(G)$ is equal to the number of prime factors of the order of G.

Several characterizations of a super-solvable group have been obtained. (Cf. ORE [*3*], ZAPPA [*1*], IWASAWA [*1*], ITÔ [*1*] and recent work by HUPPERT.) The commutator subgroup of a super-solvable group is always nilpotent.

A group is called an *L M-group* if the subgroup lattice is *lower semi-modular*, i. e. if the intersection of two subgroups U and V is maximal in V whenever U is a maximal subgroup of $U \cup V$. An $L\,M$-group of finite order is always a J-group, and so super-solvable.

Theorem 10. *A finite group G is an L M-group if and only if G is super-solvable and induces an automorphism of prime order in each factor group of a principal series* (JONES [*2*] and ITÔ [*1*]).

Proof. Suppose that G is an $L\,M$-group. By theorem 9, G is super-solvable so that G has a principal series

$$e = N_0 \subset N_1 \subset \cdots \subset N_r = G,$$

such that each group N_i/N_{i-1} $(i = 1, 2, \ldots, r)$ is of prime order. Suppose that the theorem is true for groups of smaller order. It is sufficient to prove that G induces an automorphism of prime order (or an identical one) in $N = N_1$. Let the order of N be p. G contains a p-SYLOW complement H. Consider the subgroup $K = H\,N$ and the centralizer Z of N in K. Z is the direct product of N and $M = Z \cap H$, and K/Z is isomorphic to the group of automorphisms of N induced by elements in G. In particular K/Z is cyclic. If H is a normal subgroup of K, Z coincides with K and G induces an identical automorphism in N. If H is not normal, we take a subgroup $L \neq H$ of K, conjugate to H. H and L are maximal in K, so that the intersection $H \cap L$ is a maximal subgroup of H by lower semi-modularity. Since $H \cap L \supseteq H \cap Z$, $H \cap L$ is normal and hence coincides with $H \cap Z$. This proves that $K/Z \cong H/H \cap Z$ is of prime order.

Now suppose that, conversely, G satisfies the conditions of this theorem. Using again induction on the order of G we need only to show that the intersection of two maximal subgroups M and L is maximal in M. If L contains a proper normal subgroup of G, $M \cap L$ is a maximal subgroup of M. We may, therefore, assume that both M and L

contain no proper normal subgroup of G. Take a p-SYLOW subgroup S of G which belongs to the largest prime factor of the order of G. Then $S \cap M$ is a normal subgroup of G, so that $S \cap M = e$ by our assumption. Hence S is of order p. Since $S \cap M = S \cap L = e$, M and L are conjugate to each other. Now consider the centralizer Z of S in G. Then by assumption $(G:Z)$ is a prime number. Z is a direct product of S and $Z \cap M$. Since L is conjugate to M, L contains $Z \cap M$ or $Z \cap L = Z \cap M = M \cap L$. Hence $M \cap L$ is maximal in M.

As a corollary we have

Theorem 11. *If G and H are $L\,M$-groups of finite order, then their direct product $G \times H$ is also an $L\,M$-group* (JONES [*2*], ITÔ [*1*]).

From theorem 10 we see that *a nilpotent group is always an $L\,M$-group, and so is a group of order $p^n\,q$* ($p \equiv 1 \pmod{q}$).

A finite group P is called a *P-group* if it is either

(1) an elementary abelian p-group, or

(2) a group generated by $a_1, \ldots, a_n$ and b with relations

$$a_i^p = b^q = 1,\; a_i\,a_j = a_j\,a_i \text{ and } b\,a_i\,b^{-1} = a_i^r,$$

where $r \not\equiv 1$, $r^q \equiv 1 \pmod{p}$.

The subgroup lattices of P-groups are *irreducible, complemented modular lattices*, and *groups of type* (2) *have the same lattice as elementary abelian p-groups* (BAER [*6*], IWASAWA [*1*], SUZUKI [*1*]).

Proposition 1.4. *Let G be a finite $L\,M$-group. If all intervals in $L(G)$ are irreducible and if $\Phi(G) = e$* (in particular *if $L(G)$ is a projective geometry*), *then G is a P-group.*

Proof. The commutator subgroup $C(G)$ of G is nilpotent. Hence by the irreducibility of intervals, $C(G)$ is a p-group. Here p is the largest prime factor of the order of G, as is seen from the proof of theorem 9. If G is a p-group, G must be an abelian P-group because $\Phi(G) = e$. If G is not a p-group, $G/C(G)$ is a q-group where $p > q$. Hence the order of G is $p^m\,q^n$. Consider a q-SYLOW subgroup Q of G and a subgroup H covering Q. Since the interval H/e is irreducible, Q is not normal in H and consequently there exists a subgroup $Q' \neq Q$, conjugate to Q in H. Clearly $H = Q \cup Q'$ and hence $Q \cap Q'$ is maximal in both Q and Q'. Hence $Q \cap Q'$ is normal in H and this implies $Q \cap Q' = e$ by the assumption of irreducibility of intervals.

Now $S = C(G)$ is a p-SYLOW subgroup of G. If we take a maximal subgroup M of G, $M \cap S$ is either S or maximal in S. Hence we have $\Phi(S) = e$, which implies that S is an abelian P-group. S is now a direct product of simple Q-modules S_i:

$$S = S_1 \times S_2 \times \cdots \times S_n.$$

Each S_i is then a minimal normal subgroup of G, and therefore is cyclic by theorem 9. Put $S_i = \{a_i\}$ $(i = 1, 2, \ldots, n)$ and $Q = \{b\}$. We have

$$b\, a_i\, b^{-1} = a_i^{r_i}, \text{ where } r_i^q \equiv 1 \pmod{p}.$$

Since every interval is irreducible, $r_i \not\equiv 1 \pmod{p}$ and $r_i \equiv r \pmod{p}$ for every i. G is thus a P-group.

A finite group is called a *P^*-group*, if G contains a normal subgroup N such that N is an abelian p-group, G/N is a cyclic q-group and the automorphisms of N induced by elements of G have the form: $a \to a^r$ for every $a \in N$ and r independent of a, satisfying $r \not\equiv 1$, $r^q \equiv 1 \pmod{p}$. If, moreover, N is an abelian P-group, then G is called a *P_0^*-group*. On the other hand, if the index of N is q, then G is called a *P_1^*-group*.

Proposition 1.5. *A group G is a finite $L\,M$-group and all intervals in $L(G)$ are irreducible if and only if G is either a p-group or a P_1^*-group.*

Proof. If $L(G)$ satisfies the above conditions, the order of G is either a power of p or $p^n\, q$ $(p > q)$. Suppose that G is not a p-group. Then the p-Sylow subgroup S of G is maximal. If $\Phi(G) \neq \Phi(S)$, then there would exist a normal subgroup T such that $S = T \cup \Phi(G)$, and G/T is of order $p\, q$. This is a contradiction since $\Phi(G/T) = e$, or $T \supseteq \Phi(G)$. Hence we must have $\Phi(G) = \Phi(S)$.

We use induction on n. Suppose that (1.5) is proved for all maximal subgroups of G. Take a maximal subgroup U of S. Since $\Phi(G) = \Phi(S)$ and $G/\Phi(G)$ is a P-group by (1.4), there exist two distinct maximal subgroups M_1 and M_2 of G such that $U = S \cap M_1 = S \cap M_2$. By inductive hypothesis both M_1 and M_2 are P_1^*-groups and therefore all subgroups of U are normal in both M_1 and M_2 and therefore in G. This implies that S is an abelian group, since $p > q \geq 2$, and that G is a P_1^*-group. The validity of the converse is fairly obvious.

Now we obtain the following theorem (Suzuki [*1*], p. 352, theorem 3).

Theorem 12. *Let G be a finite p-group. If there is an isomorphic mapping from $L(G)$ onto a subgroup lattice $L(H)$ of another group H, H is also a p-group, except when G is either a cyclic group or an abelian P-group.*

Proof. By assumption H is an $L\,M$-group and all intervals in $L(H)$ are irreducible. Suppose now that H is not a p-group. If H is of prime power order, G must be cyclic, since otherwise p can be determined by the lattice of subgroups of H. If H is not of prime power order, H is a P_1^*-group. Suppose S contains an element a of order p^2. Then H contains a subgroup K which is a P_1^*-group and the p-Sylow subgroup of K is a cyclic group of order p^2. K contains only one chain of dimension 2, $1 + p$ maximal subgroups and $1 + p^2$ minimal subgroups. Since $p > 2$, this group can not be the image of a p-group under any lattice isomorphism. Hence all elements of S are of order p, and this proves that H is a P-group. Hence G must be an abelian P-group.

4. Finite groups with a modular lattice of subgroups.

A group G is called a *modular group* or an *M-group* if the subgroup lattice $L(G)$ of G is modular. The structure of finite M-groups has been determined by IWASAWA [*1*], and supplemented by JONES [*1*].

Proposition 1.6. *If G is a finite M-group such that $G/\Phi(G)$ is a non-abelian P-group, then G is a P_0^*-group.*

Proof. By assumption, the p-SYLOW subgroup S of G is normal and G contains only one subgroup of index q. Hence a q-SYLOW subgroup Q of G is cyclic: $Q = \{b\}$. Take an element a of order p. Then Q is a maximal subgroup of $\{Q, a\}$. Hence $b\,a\,b^{-1} = a^r$, and r does not depend on the choice of a. Hence by assumption $r \not\equiv 1 \pmod p$ and the normalizer of Q coincides with Q. Now $\Phi = \Phi(G)$ is nilpotent and is a direct product of $S \cap \Phi$ and $Q \cap \Phi$.

For any element $c \in S$, $Q \cap c\,Q\,c^{-1} \supseteq Q \cap \Phi$. Hence if $c \neq 1$, $Q \cap c\,Q\,c^{-1} = Q \cap \Phi$. By modularity $Q \cup c\,Q\,c^{-1}$ is a group of order $p\,q^n$ and there is an element a of order p such that $c\,Q\,c^{-1} = a\,Q\,a^{-1}$. Since the normalizer of Q is Q, we have $a = c$, which implies that S is an abelian P-group. Hence G is a P_0^*-group.

Theorem 13. *A finite M-group G is a direct product of groups $P_1, \ldots, P_n$ such that each P_i is either a modular p-group or a P_0^*-group, and P_i and P_j $(i \neq j)$ have relatively prime orders* (IWASAWA [*1*], p. 181, Satz 3).

Proof. Let Φ be the Φ-subgroup of G. Then $L(G/\Phi)$ is a complemented modular lattice. Hence by a well-known theorem of lattice theory $L(G/\Phi)$ is decomposable into a direct product of irreducible components. Hence by theorem 4, p. 5, G/Φ is a direct product of P-groups. In virtue of (1.1) of p. 5 and (1.6), G is a direct product of p-groups and P_0^*-groups.

It is consequently sufficient to determine the structure of *modular p-groups*. This has been done by IWASAWA [*1*].

Theorem 14. *A non-Hamiltonian p-group G is modular, if and only if G contains an abelian normal subgroup N with cyclic factor group G/N and there exist an element t in G with $G = \{N, t\}$ and an integer s which is at least 2 in case $p = 2$ such that $t\,a\,t^{-1} = a^{1+p^s}$ for every a in N.*

The proof of this theorem is not simple. We shall begin with a series of preliminary propositions.

Proposition 1.7. *Let G be a modular p-group. The totality of elements of order p forms an abelian characteristic subgroup of G.*

Proof. It suffices to prove that two elements of order p commute. If a and b are two elements of order p, then the subgroup $\{a, b\}$ generated by a and b is of order p^2, since the subgroup lattice of $\{a, b\}$ is of dimension 2. Hence $\{a, b\}$ is abelian.

In virtue of (1.7), *the totality of elements of order less than or equal to p^α forms a characteristic subgroup of G, which is denoted by Ω_α.*

Proposition 1.8. *Let G be a p-group. If G is not an M-group, but all proper subgroups are M-groups, then G contains a normal subgroup N such that the factor group G/N is of order p^3 and not an M-group.*

Proof. By assumption G contains subgroups U and V such that

$$U \cup V = G,\ (U: U \cap V) = (V: U \cap V) = p,\ (G: U) \geq p^2 \text{ and } (G: V) \geq p^2.$$

Since $U \cap V$ is a normal subgroup of G, we may assume that $U \cap V = e$. By this assumption, $U = \{u\}$, and $V = \{v\}$ are both cyclic of order p. If A is the minimal normal subgroup of G containing U, then $A \cap V = e$ and $A \cup V = G$. Hence A is a maximal subgroup of G. Since A is generated by u, $v\,u\,v^{-1}, \ldots$, A is by (1.7) an elementary abelian group. A contains a normal subgroup D of G such that $(A:D) = p$. By assumption $\{D, v\}$ is an M-group so that by (1.7) v commutes with every element of D. This implies that D is contained in the center of G, and in particular the commutator $w = v\,u\,v^{-1}\,u^{-1}$ is a central element of order p. Hence u and v generate a p-group of order p^3 which is not an M-group.

The structure of groups of order p^4 has been completely determined (Cf. BURNSIDE: *Theory of groups*, p. 126). We can easily see that every non-abelian M-group of order p^4 belongs to one of the following types:

(1) $u^{p^3} = v^p = 1,\ v\,u\,v^{-1} = u^{1+p^2}$,

(2) $u^{p^2} = v^p = w^p = 1,\ v\,u\,v^{-1} = u^{1+p},\ v\,w = w\,v,\ u\,w = w\,u$ $(p > 2)$,

(3) $u^{p^2} = v^{p^2} = 1,\ v\,u\,v^{-1} = u^{1+p}$ $(p > 2)$,

(4) $u^4 = w^2 = 1,\ u^2 = v^2,\ v\,u\,v^{-1} = u^{-1},\ w\,u = u\,w,\ v\,w = w\,v$.

It is easily proved that $(a\,b)^p = a^p\,b^p$ for any two elements a and b of a group of type (2) or (3).

Let G be a modular p-group and let p^μ be the maximal order of elements in G. $\mu = \mu(G)$ is an invariant of G.

Assume now that G is a 2-group and $\mu(G) = 2$. If G does not contain a quaternion group, any pair of elements is permutable so that G is abelian. Suppose G contains a quaternion group $Q = \{u, v\}$, $u^4 = 1$, $v^2 = u^2$, $v\,u\,v^{-1} = u^{-1}$. If $G \neq Q\,\Omega_1$, we take an element $w \notin Q\,\Omega_1$. If $w^2 = u^2$, the subgroup $\{u, v, w\}$ is not modular, contrary to the assumption. If $\{w\} \cap Q = e$, w commutes with u and v. Hence the subgroup $\{u, v, w\}$ is again not modular. We have, therefore, $G = \Omega_1 Q$. Let

$$\Omega_1 = \{u^2\} \times \{c_1\} \times \cdots \times \{c_n\},\ c_i^2 = 1.$$

Then each c_i commutes with both u and v. Hence $G = T \times Q$ and G is a Hamiltonian 2-group. Suppose that G is a general modular 2-group. Each factor group Ω_i/Ω_{i-2} is, as shown above, abelian or Hamiltonian. Suppose that one of the Ω_i/Ω_{i-2} is not abelian. Then Ω_i/Ω_{i-2} contains a quaternion group $Q = \{u, v\}$. If $i < \mu(G)$, there is an element x of G such that $\bar{x}$, the coset modulo Ω_{i-2}, satisfies the relation $\bar{x}^4 = u^2 = v^2$. $\{\bar{x}, u, v\}$ is then not modular as is shown by simple computation. Hence $i = \mu$. If $i > 2$, we have again a contradiction. Hence if G is not Hamiltonian, each factor group Ω_i/Ω_{i-2} is abelian.

From the above consideration we can prove that, if a modular p-group is not Hamiltonian, then $(u\,v)^{p^{\mu-1}} = u^{p^{\mu-1}}\,v^{p^{\mu-1}}$ for any pair of elements u, v. This is fairly easy for modular groups with $\mu(G) = 2$, since every pair of elements generates an abelian subgroup or a group of type (3) or (4). The general case follows by induction.

Let G be again a modular p-group which is not Hamiltonian. Denote the index $(\Omega_\alpha : \Omega_{\alpha-1}) = p^{\omega_\alpha}$ for each $\alpha = 1, 2, \ldots, \mu$. We can take elements $a_1, \ldots, a_{\omega_\mu}$ of G, so that they form a basis for $\Omega_\mu/\Omega_{\mu-1}$. Now $a_1^p, \ldots, a_{\omega_\mu}^p$ can be extended to the basis $a_1^p, \ldots, a_{\omega_\mu}^p, a_{\omega_\mu+1}, \ldots, a_{\omega_{\mu-1}}$ of $\Omega_{\mu-1}/\Omega_{\mu-2}$, and so on. Since G is quasi-Hamiltonian by theorem 7, every element of G can be written as a product of powers of these $a_1, \ldots$, in other words: the elements $a_1, \ldots$ form a *basis* of G.

Now we can prove the following

Proposition 1.9. *Let G be a modular p-group which is not Hamiltonian. If the commutator subgroup $C(G)$ is of order p, then*

$$G = G_1 \times G_2$$

where $G_1 = \{a_1, a_2\}$, $a_1^{p^m} = a_2^{p^n} = 1$, $a_2\,a_1\,a^{-1} = a_1^{1+p^{m-1}}$, G_2 is abelian and $G_2^{p^{m-1}} = e$.

Proof. Since G has a basis $a_1, a_2, \ldots, a_s$, we may assume that $C(G)$ is contained in $\{a_1\}$. By the modularity, we have $\{a_i, a_j\} \cap C(G) = e$ $(i, j > 1)$, which implies that $\{a_2, \ldots, a_s\}$ is abelian. Let a_2 be the element with the smallest order among those basis elements $a_2, \ldots, a_s$ which do not commute with a_1. Taking suitable basis elements, we may assume that $G = G_1 \times G_2$, where $G_1 = \{a_1, a_2\}$ and $G_2 = \{a_3, \ldots, a_s\}$. Now

$a_2 a_1 a_2^{-1} = a_1^{1+p^{m-1}}$ where $a_1^{p^m} = 1$. Since

$$a_2 (a_1 a_i) a_2^{-1} = a_1^{1+p^{m-1}} a_i = (a_1 a_i)^x \quad (i > 2),$$

the order of a_i is at most p^{m-1}.

Proof of theorem 14. Suppose first that G has the structure described in theorem 14. Then all subgroups and factor groups of G satisfy the same conditions as G. Hence using induction on the order of G, we conclude from (1.8) that G is modular.

Now suppose conversely G to be an M-group and not Hamiltonian. Using again induction, we assume this theorem to be true for groups of smaller order. As remarked before, the mapping $g \to g^{p^{\mu-1}}$ $(\mu = \mu(G))$ is a homomorphism of G. Denote the image of this homomorphism by $\mho$. It is obvious that $\mho$ is a normal subgroup of G. Hence by a property of p-groups there exists a normal subgroup Z of order p contained in $\mho$. By inductive hypothesis, G/Z contains an abelian normal subgroup N_1/Z with properties stated in theorem 14. Let t be an element of G such that $G = \{N_1, t\}$ and $t a t^{-1} = a^{1+p^s} z^k$ for any element $a \in N_1$.

(I) Suppose N_1 to be *abelian*. If $\{t\} \cap Z = e$, we take a basis $a_1, a_2, \ldots, a_r$ of N_1 such that $Z \subseteq \{a_1\}$ and $N_1 \cap \{t\} \subseteq \{a_2\}$. For $i \geq 2$, $\{a_i, t\} \cap Z = e$ implies that

$$t a_i t^{-1} = a_i^{1+p^s} \quad (i \geq 2).$$

For $i = 1$, we have

$$t a_1 t^{-1} = a_1^{1+p^s+kp^{m-1}},$$

where p^m is the order of a_1. If the order of a_i $(i \geq 3)$ is larger than p^m,

$$t (a_1 a_i) t^{-1} = a_1^{1+p^s+kp^{m-1}} a_i^{1+p^s} = (a_1 a_i)^x = a_1^x a_i^x,$$

which implies that $x \equiv 1 + p^s \pmod{p^m}$, or $k \equiv 0 \pmod{p}$. Hence we have

$$t a_1 t^{-1} = a_1^{1+p^s}.$$

If the order of a_2 is greater than p^m, we can replace t by $t^* = t a_2$ so that $\{a_2\} \cap \{t^*\} = e$. Then by a similar computation we get

$$t^* a_i t^{*-1} = a_i^{1+p^s}.$$

Suppose that the order of a_i $(i \geq 2)$ is smaller than p^m. If $m - 1 \leq s$, G is either abelian or the commutator subgroup of G is of order p. This case has been treated in (1.9). If $m - 1 > s$, we take a power $t^* = t^x$, where $x = 1 - k p^{m-1-s}$. It is now easy to show that

$$t^* a_i t^{*-1} = a_i^{1+p^s} \text{ for all } i.$$

Suppose next $\{t\} \supseteq Z$. We take a basis $a_1, \ldots, a_r$ of N_1 such that

$$\{t\} \cap N_1 \subseteq \{a_1\}.$$

Let the order of t be p^n. If some a_i has an order not smaller than p^n, then we can replace t by $t\, a_i^x$ and reduce this case to the above. Hence we may assume that $n = \mu$ and $a^{p^{\mu-1}} = 1$ for all i. In this case we may assume moreover, without loss of generality, that

$$t\, a_i\, t^{-1} = a_i^{1+p^s} \text{ for } i \geq 3,$$

and

$$t\, a_2\, t^{-1} = a_2^{1+p^s} \text{ or } a_2^{1+p^s}\, t^{p^{\mu-1}}.$$

If

$$t\, a_1\, t^{-1} = a_1^{1+p^s+h\,p^{m-1}} \text{ and } t\, a_i\, t^{-1} = a_i^{1+p^s} \quad (i \geq 2),$$

we may replace a_1 by $a_1^* = a_1\, t^\alpha$, where $\alpha = k\, p^{\mu-1-s}$ and $a_1^{h p^{m-1}} = t^{k p^{\mu-1}}$. It can be seen that $a_1^*, a_2, \ldots, a_s$ form an abelian normal subgroup of G and

$$t\, a_1^*\, t^{-1} = a_1^{*1+p^s}, \text{ and } t\, a_i\, t^{-1} = a_i^{1+p^s} \quad (i \geq 2).$$

If

$$t\, a_2\, t^{-1} = a_2^{1+p^s}\, t^{p^{\mu-1}},$$

and if $s < \mu - 1$, we replace a_2 by $a_2^* = a_2\, t^{p^{\mu-1-s}}$ and reduce this case to the above proved one. If $s \geq \mu - 1$, G is either an abelian group, or the commutator subgroup of G coincides with Z. Hence this theorem is a consequence of (1.9).

(II) Suppose N_1 is *not* abelian. By (1.9), N_1 has the following structure:

$$N_1 = \{a_1, a_2\} \times \{a_3, \ldots, a_r\},$$

$$a_2\, a_1\, a_2^{-1} = a_1^{1+p^{m-1}}, \; a_i\, a_j = a_j\, a_i \quad (i, j \geq 3).$$

We may assume that

$$t\, a_1\, t^{-1} = a_1^{1+p^s}, \text{ and } t\, a_2\, t^{-1} = a_2^{1+p^s},$$

replacing t by a suitable $t\, a_1^\alpha\, a_2^\beta$. Since $Z \subseteq \mho$, the element z of Z is a $p^{\mu-1}$-th power of some element. Suppose

$$a_1^{p^{m-1}} = c^{p^{\mu-1}} \text{ and } c = a_1^{e_1}\, a_2^{e_2} \cdots a_r^{e_r}\, t^f.$$

Then

$$a_1^{p^{m-1}} = c^{p^{\mu-1}} = a_1^{e_1 p^{\mu-1}}\, a_2^{e_2 p^{\mu-1}}\, t^{f p^{\mu-1}}.$$

If $a_1^{p^{\mu-1}} \neq 1$, then $\mu = m$. If $s = \mu - 1$ and $a_2^{\mu^{p-1}} = 1$, then the commutator subgroup of G coincides with the center and our argument is completed. If $a^{p^{\mu-1}} \neq 1$, we replace a_2 by $a_2^* = a_2 t^{-1}$, and reduce this case to the former one. If $s < \mu - 1$, we can again reduce it to the case (I) replacing a_2 by $a_2^* = a_2 t^{-p^{\mu-1-s}}$.

Suppose finally that $a_1^{p^{\mu-1}} = 1$. Then $t^{p^{\mu-1}} \neq 1$; otherwise we would have $a_1^{p^{m-1}} = a_2^{e_s p^{\mu-1}}$. Denote by p^l the index

$$(\{a_1, a_2, t\} : \{a_1, a_2\}),$$

and put

$$t^{\mu^l} = a_1^{-p^k} a_2^{-h p^f}.$$

Then by assumption $l > k$. If $k \geq 1$, we take

$$g = t^{p^{l-1}} a_1^{x p^{k-1}} a_2^{y p^{f-1}} \text{ and } g^* = t^{p^{l-k}} a_1^z$$

such that

$$g^p = 1 \text{ and } g^{*p^{k-1}} = t^{p^{l-1}} a^{x p^{k-1}}.$$

Then g and a suitable power of a_2 generate Ω_1 in $H = \{g^*, a_2\}$. Since

$$a_2 g^* a_2^{-1} = a_2^u g^* a_1^{z p^{m-1}},$$

we have

$$a_1^{p^{m-1}} \in \{a_2^{p^t}, g\}$$

which is a contradiction. Hence $k = 0$ and $G = \{t, a_1, \ldots, a_r\}$. This reduces the case (II) to the first case.

As a corollary we have the following theorem.

Theorem 15. *Finite M-groups are always metabelian.*

5. Structure of infinite M-groups.

Infinite M-groups are considered by IWASAWA [2], BEAUMONT [2] and ZAPPA [2, 3, 4, 5]. IWASAWA determined almost completely the structure of such groups.

Proposition 1.10. *The set $E = E(G)$ of elements of finite order in an M-group G is a characteristic subgroup of G* (ZAPPA [2], IWASAWA [2]).

Proof. Let a and b be two elements of finite order in G. Then the dimension of $L\{a, b\}$ is finite, which implies that every element in $\{a, b\}$, in particular $a b$ is of finite order.

First of all consider an M-group, which contains at least one element of infinite order. Let G be an M-group containing an element u of infinite order. If we take another element v of G such that

$$\{u\} \frown \{v\} = e,$$

then the interval $\{u, v\}/\{v\}$ is isomorphic to $\{u\}/e$. Hence the subgroups

$$\{u^i\} \smile \{v\} \quad (i = 1, 2, \ldots)$$

are all different from each other and exhaust all subgroups between $\{u, v\}$ and $\{v\}$. On the other hand, if v is of finite order, $\{v, u\,v\,u^{-1}\}$ contains no element of finite order by (1.10). Hence if v is of finite order, and if u is of infinite order, then

$$u\,v\,u^{-1} = v^r.$$

Proposition 1.11. *If an M-group contains an element of infinite order, the subgroup E consisting of all elements of finite order is abelian* (IWASAWA [2], p. 719, theorem 5).

Proof. Let u be an element of infinite order, and let a and b be any two elements of finite order. Then au has an infinite order by (1.10). Hence

$$u\,b\,u^{-1} = b^s \text{ and } (a\,u)\,b\,(a\,u)^{-1} = b^t.$$

Consequently

$$a\,b^s\,a^{-1} = b^t.$$

Since both s and t are relatively prime to the order of b, E is either abelian or Hamiltonian. If E is Hamiltonian, E contains a quaternion group Q. u induces an automorphism in Q. A suitable power u_0 of u is therefore commutative with all elements in Q. Hence $\{u_0, Q\}/\{u_0^4\}$ is not an M-group (cf. § 4). This shows that E is abelian.

Now we prove

Proposition 1.12. *For any M-group G, G/E is abelian* (IWASAWA [2], p. 720, theorem 6).

Proof. Let G be an M-group which contains no element of finite order except the identity. Take two elements u and v of G, and suppose

$$\{u\} \frown \{v\} = \{w\}, \quad w = u^\alpha \neq 1.$$

By way of contradiction, suppose $u \neq v\,u\,v^{-1}$. Since $u^\alpha = v\,u^\alpha\,v^{-1}$, we have

$$\{u\} \frown \{v\,u\,v^{-1}\} = \{z\}, \quad z = u^\beta = v\,u^\gamma\,v^{-1} \neq 1.$$

Since u is of infinite order, $\beta = \pm\gamma$. But if $\gamma = -\beta$, $\{u^\beta, v\}/\{u^{4\beta}, v^2\}$ is not an M-group. Hence $\beta = \gamma$. $\{u, v, u\,v^{-1}\}/\{z\}$ is an M-group generated by two elements of finite order, so that $(u\,v\,u^{-1}\,v^{-1})^\delta = z^\varepsilon$. Then the

2*

modular identity is not satisfied by subgroups

$$A = \{u\, z^s\},\ B = \{v\, u\, v^{-1}\, z^s\} \text{ and } C = \{u\}.$$

Hence we have $u = v\, u\, v^{-1}$ which implies that $\{u, v\}$ is cyclic.

Suppose next that $\{u\} \cap \{v\} = e$. If we have $u^\alpha\, v^\beta = v^\beta\, u^\alpha$ for some positive integers α and β, then $\{u\} \cap \{v^\beta\, u\, v^{-\beta}\} \neq e$. As before we must have $u = v^\beta\, u\, v^{-\beta}$. Similarly we have $u\, v = v\, u$. Let

$$H = \{u, v\},\ U = \{u\, v\, u^{-1}, v\},\ V = \{u, v^2\} \text{ and } W = \{u\, v^2\, u^{-1}, v^2\}.$$

Then $U = \{u^\alpha, v\}$ or $U = \{v\}$. Clearly U is a normal subgroup of H, and similarly W is a normal subgroup of V. Now $W = \{(u\, v) v^2\, (u\, v)^{-1}, v^2\}$, so that W is self-conjugate in $\{u\, v, v^2\}$. This implies that W is a normal subgroup of H. U/W is then generated by two elements of order 2, and consequently U/W is a finite group. Assume first that H/W is infinite. Then H/U is also infinite and $U = \{v\}$. Hence $u\, v\, u^{-1} = v^\beta$ which implies that $u\, v = v\, u$. Next let H/W be finite. Since $V \supseteq W$ and $H \neq V$, we have $H \neq W$. Hence the commutator subgroup $C = C(H)$ of H does not coincide with H. If $C \subseteq \{u\}$ and $\{v\}$, $C \subseteq \{u\} \cap \{v\} = e$ which implies that H is abelian. We may therefore assume for instance

$$C \subsetneq \{u\}, \text{ or } C \cup \{u\} \neq \{u\}.$$

Hence $C \cup \{u\} = \{u^\beta, v\}$ $(\beta > 0)$. Since $C \cup \{u\}/C$ is cyclic, there exists a γ such that $u^\gamma \equiv v^\beta \pmod{C}$. $w = u^\gamma\, v^{-\beta} \in C$ and

$$\{u\} \cup C = \{u, v^\beta\} = \{u, w\}.$$

If $\{u\} \cap \{w\} \neq e$, $\{u\} \cup C$ is abelian. Hence $uv^\beta = v^\beta u$ and H is abelian as before. If $\{u\} \cap \{w\} = e$, we have

$$C = \{u^n, w\}, \text{ or } C = \{w\}.$$

If $C = \{w\}$, then $u\, w = w\, u$, or $u\, v^\beta = v^\beta\, u$. This shows that $u\, v = v\, u$. If $C = \{u^n, w\}$, H/C is a finite group. In this case C is not abelian; otherwise H would be abelian and $C = e$. Hence C has the same structure as H. This shows that the commutator subgroup C_1 of C is also not the unit group, and C/C_1 is finite. Similarly if we consider the commutator subgroup C_2 of C_1, C_1/C_2 is also finite, so that H/C_2 is a finite M-group. This is, however, impossible according to theorem 15. Hence H must be abelian in all cases.

The last half of the above proof shows the validity of the following proposition.

Proposition 1.13. *Let G be an M-group generated by two elements u and v of infinite order such that $\{u\} \cap \{v\} = e$. Then G is a torsion-free abelian group.*

Theorem 16. *Let G be an M-group. If G contains two elements u and v of infinite order such that $\{u\} \cap \{v\} = e$, then G is abelian* (IWASAWA [2], p. 724, theorem 7).

Proof. If we take two arbitrary elements x and y of infinite order, then

$$\{x\} \frown \{u\} = e \text{ or } \{x\} \frown \{v\} = e.$$

Suppose for instance $\{x\} \frown \{u\} = e$. If $\{x\} \frown \{y\} = e$, then $x\,y = y\,x$ by (1.13). If $\{x\} \frown \{y\} \neq e$, then $\{x\} \frown \{u\,y\} = e$, and $u\,y$ is of infinite order by (1.13). Hence $(u\,y)\,x = x\,(u\,y)$. Since $x\,u = u\,x$ by (1.13), we have again $x\,y = y\,x$. Hence elements of infinite order are commutative with another. Let now z be an element of finite order. Then $u\,z$ is of infinite order by (1.10). Hence $(u\,z)\,x = x\,(u\,z)$, which implies $z\,x = x\,z$.

Theorem 17. (IWASAWA [2], p. 727, theorem 8.) *Let G be an M-group and let E be the normal subgroup of G consisting of all elements of finite order in G. If the abelian group G/E is of rank 1, then G has the following structure: $G = \{E, z_1, z_2, \ldots\}$ where z_1 is of infinite order, $z_{i-1}^{p_i} = z_i\,e_i$, $z_{i-1}\,e_i\,z_{i-1}^{-1} = e_i^{\beta_i}$ (p_i a prime number, $e_i \in E$), and for any element a of the p-component E_p of E, $z_i\,a\,z_i^{-1} = a^{\alpha_i(p)}$ where $\alpha_i(p)$ is a p-adic number, uniquely determined modulo the exponent p^n of E_p and*

$\alpha_i(p) \equiv 1 \pmod p$ $\bigl(\alpha_i(2) \equiv 1 \pmod 4\bigr)$, $\alpha_{i-1}^{p_i}(p) \equiv \alpha_i(p) \pmod{p^n}$.

Proof. First of all, assume that G/E is an infinite cyclic group. Let z be one of the generators of G modulo E. Let E_p be the p-component of E and P_k be the subgroup of E_p consisting of all elements which satisfy the relation $x^{p^k} = 1$. Since P_k has a basis, we can easily prove that $z\,a\,z^{-1} = a^{r_k}$ for all $a \in P_k$. We have now $r_k \equiv r_{k-1} \pmod{p^k}$, so that $\{r_k\}$ has a p-adic limit $\alpha(p)$, and $\alpha(p)$ is determined uniquely modulo the exponent of E_p. If $r_1 \not\equiv 1 \pmod p$, we may assume, taking a suitable power u of z, that

$$u\,a\,u^{-1} = a^t,\ t \not\equiv 1,\ t^q \equiv 1 \pmod p$$

for an element a of order p. It is easy to show that $\{a, u\}/\{u^{p\,q}\}$ is not an M-group. Hence $\alpha(p) \equiv 1 \pmod p$. Similarly we prove that $\alpha(2) \equiv 1 \pmod 4$. The rest of our theorem can be verified easily.

Conversely the groups having the structure of theorem 17 are always M-groups. In fact it can be shown that any group having the above structure is quasi-Hamiltonian (IWASAWA [2]), and hence modular.

The structure of infinite M-groups without elements of infinite order is not yet known. IWASAWA [2] determined the structure of such groups under the assumption that the groups involved are *locally finite.* Here we call a group locally finite, if every finite set of elements generates a finite subgroup. Strictly speaking, we need not assume this strong condition. If an M-group G is generated by two elements of finite order, then the subgroup lattice of G has a finite dimension in virtue of the modular identity. The following argument is true for a class of torsion groups which satisfy the condition: *every subgroup with a finite dimensional lattice is finite.* It is still an open question whether M-groups are in this class of groups. A stronger statement has been conjectured by SCHUR: namely if a group satisfies both maximal and minimal conditions for subgroups, then this group would be finite. It is still an open question whether this conjecture is true in general, but it is true for the class of groups with isomorphic representations by matrices. We prove the following theorem of IWASAWA ([2], theorems 2 and 3).

Theorem 18. *Let G be a locally finite M-group. Then G is a direct product of p-groups and groups $\{P, u\}$, where P is an abelian p-group, u is an element of order q^n $(p > q)$, and*

$$x^p = 1,\ u\,x\,u^{-1} = x^r,\ r \not\equiv 1,\ r^q \equiv 1 \pmod p$$

for every x in P.

A locally finite non-abelian p-group G is an M-group if and only if G is either a Hamiltonian group or G contains an abelian normal subgroup A with the following properties:

(1) *the orders of the elements of A are bounded* (we denote the maximal order by p^n),

(2) *G/A is a cyclic group of order p^m,*

(3) *there exist an element t of G and an integer s, such that*

$$G = \{A, t\} \text{ and } t\,a\,t^{-1} = a^{1+p^s} \text{ for all } a \in A,$$

$$n \leq s + m,\ t^{p^{s+m}} = 1 \text{ and if } p = 2,\ s \geq 2.$$

Proof. Let G be a locally finite M-group. Assume that two elements u and v of q-power order generate a non-nilpotent subgroup of G. Then $\{u, v\}$ is a finite group by assumption and has an order $p^\alpha q^\beta$ by theorem 13. Moreover, a p-SYLOW subgroup of $\{u, v\}$ is an abelian P-group by theorem 13. Let x be an element of G with p-power order. $\{x, u, v\}$ is, then, a non-nilpotent M-group of order $p^\gamma q^\beta$, so that $x^p = 1$ and this implies that all elements of G with p-power order form a characteristic abelian subgroup P satisfying $P^p = e$. Let A be the p-SYLOW subgroup and $\{w\}$ one of the q-SYLOW subgroups of $\{x, u, v\}$. It holds now that

$$w\,x\,w^{-1} = x^r,\ r \not\equiv 1,\ r^q \equiv 1 \pmod p.$$

Suppose $w^\alpha \equiv u \pmod A$, then

$$u\,x\,u^{-1} = x^s,\ s \not\equiv 1 \pmod p.$$

We have, therefore,

$$u\,x\,u^{-1} = w^\alpha\,x\,w^{-\alpha} = x^{r^\alpha} = x^s, \text{ or } r^\alpha \equiv s \not\equiv 1 \pmod p.$$

Hence $\alpha \not\equiv 0 \pmod{q}$. This means that $\{u\}$ is also a q-SYLOW subgroup of $\{x, u, v\}$; every element of q-power order is contained in $\{P, u\}$. This proves the first part of this theorem.

If a locally finite p-group G is modular, and if G contains a quaternion group, then it is easily shown that G is Hamiltonian. We consider finally the case in which the modular p-group G is neither abelian nor Hamiltonian.

In this case G contains a non-abelian finite subgroup H which has the structure described in theorem 14. Hence H contains an abelian normal subgroup A such that H/A is cyclic and $t\,a\,t^{-1} = a^{1+p^s}$ for all $a \in A$ and a generator t of G modulo A. We may write $H = \{A, t; s\}$. This representation of H may not be unique. We denote by $s(H)$ the maximal exponent s among the different representations of H. Since H is not abelian, $s(H)$ is determined uniquely. Let s_0 be the minimum of $s(H)$ for all finite non-abelian subgroups H of G. We choose an H_0 such that $H_0 = \{A_0, t_0; s_0\}$. Take a finite subgroup K containing H_0, and let $K = \{A, t; s\}$, where $s = s(K)$. Then

$$H_0 = \{H_0 \cap A, t^*;\; s + u\} \text{ where } p^u = [K:A \cap H_0].$$

From the definition of s_0 we conclude that $s + u \leq s_0 \leq s$, and $s = s_0$, $u = 0$. This implies that

$$K = H_0 \cap A \text{ and } K = \{A, t^*;\; s\}.$$

Let the order of t^* be p^l. Then for any element a of A, we have

$$a = t^{*p^l}\, a\, t^{*-p^l} = a^{(1+p^s)\,p^l}.$$

Hence the order of a is at most p^{s+l}. This implies that the order of any element of K is at most p^{s+l}, and p^{s+l} is determined by H_0. Since K is an arbitrary finite subgroup of G containing H_0, the orders of the elements of G are bounded. Returning to K, the commutator subgroup of K is A^{p^s}, and hence the maximal order of elements in A is determined by K and does not depend on the choice of A. Denote by $O(K)$ the maximal order of elements in A. Since the orders of the elements of G are bounded, $O(K)$ is bounded, and we can choose a subgroup H such that $O(H)$ is maximal. We may assume that $H \supseteq H_0$. It can be shown that for a suitable choice of an abelian normal subgroup A^* of H, H can be expressed as $\{A^*, t; s\}$ and any finite subgroup L of G containing H is written as $\{A, t; s\}$ where $A \supseteq A^*$ and $L = A \cup H$. If we have two representations $L = \{A_1, t; s\} = \{A_2, t; s\}$, then $A_1 \cup A_2$ is proved to be abelian. For, letting $A_1 \cup A_2 = \{A_1, t^{p^u}\}$, we can conclude that $s + u \geq k$ where $p^k = O(H)$. Hence t^{p^u} commutes with every element of A_1. Hence there is a unique maximal normal subgroup A of L such that $L = \{A, t; s\}$. For arbitrary elements x, y of G we may form

$$H_x = \{H, x\} = \{A_x, t;\ s\}.$$

and similarly H_y and $H_{x,y} = \{A_{x,y}\, t;\ s\}$. In these representations we may assume A_x, A_y, $A_{x,y}$ are the unique normal subgroups considered above. It follows easily that $A_x = A_{x,y} \cap H_x$ and $A_y = A_{x,y} \cap H_y$. Since $A_{x,y}$ is abelian, $A_x \cup A_y$ is abelian, and this implies that the subgroup $N = \bigcup_{x \in G} A_x$ is abelian and that $t\,x\,t^{-1} = x^{1+p^s}$ for any $x \in N$.

The converse statement is almost clear. If we take two elements x and y of G, they generate a finite M-group by theorem 15. Hence $x\,y = y^\alpha\, x^\beta$, since a finite M-group is always quasi-Hamiltonian. Hence G is also quasi-Hamiltonian and a fortiori modular.

Any quasi-Hamiltonian group without elements of infinite order is always locally finite. Hence the above theorem 16, 17 and 18 give the structure of quasi-Hamiltonian groups. In particular we have the following theorem.

Theorem 19. *Any quasi-Hamiltonian group is metabelian.*

6. Structure of *UM*-groups.

Sato [*1, 2*] considered the structure of groups whose lattice of subgroups is *upper semi-modular*. Here a lattice is termed upper semi-modular if $a \cup b$ covers a whenever $a \cap b$ is maximal in b. A group is called an *U M-group* if its subgroup lattice is upper semi-modular. The determination of finite $U\,M$-groups was also announced by Jones [*2*].

Theorem 20. *A finite group G is an $U\,M$-group if and only if G is a direct product of groups H_i such that the orders of H_i and H_j $(i \neq j)$ are relatively prime and each group H_i is either a modular p-group or a group of the following type:*

$H = (P_1 \times \cdots \times P_r) \cup Q$ *where each P_i is a p_i-Sylow subgroup and Q is a q-Sylow subgroup, and moreover*

(1) *each P_i is elementary abelian,*

(2) *Q is cyclic: $Q = \{b\}$,*

(3) *$b\,a_i\,b^{-1} = a_i^{r_i}$ for every element $a_i \in P_i$, where $r_i \not\equiv 1$, $r_i^{q^{\beta_i}} \equiv 1$* (mod *$p$*), *and*

(4) *if β_i is chosen as small as possible in* (3), *then*

$$\beta_i \neq \beta_j \ (i \neq j).$$

Proof. Suppose that G is an $U\,M$-group of finite order. First of all we shall prove that a non-normal Sylow subgroup of G is cyclic. Now suppose that the p-Sylow subgroups of G are not normal and let D be one of the maximal intersections of two p-Sylow subgroups, say of S and T; $D = S \wedge T$. Take a subgroup H of T which covers D. By assumption $U = S \vee H$ covers S. Hence the index $(U:S)$ is a prime number because G is a J-group. It is, therefore, necessary to prove the following: if a p-Sylow subgroup S of an UM-group G of order $p^\alpha q$ (p, q are primes) is not self-conjugate, then S is cyclic. We prove this by induction on α. If $\alpha = 1$, our assertion is trivial. Suppose $\alpha = 2$. Then the centralizer of a q-Sylow subgroup is of order q or $p\,q$. In the former case, S is cyclic, as it is isomorphic with some subgroup of the group of automorphisms of a cyclic group of order q. If its order is $p\,q$, G is an LM-group by theorem 10. Hence by a theorem of lattice theory (Birkhoff's theorem) G is an M-group, so that S is cyclic. In general, let $D \subsetneq S$ be a maximal intersection of p-Sylow subgroups. Then S/D is cyclic. If S contained a maximal subgroup M, not containing D, then $D_1 = D \wedge M$ would be a normal subgroup of G. Take a subgroup U of S covering D. $U S_q/D_1$ would be an UM-group of order $p^2 q$ with non-cyclic non-normal p-Sylow subgroups. This is impossible. Thus S contains only one maximal subgroup and so S is cyclic.

We take two Sylow subgroups S_p and S_q. If the orders of S_p and S_q are p^α and q^β respectively and if $p \neq q$, $S_p \vee S_q$ is a subgroup of order $p^\alpha q^\beta$. Now suppose that a q-Sylow subgroup S_q of G is not self-conjugate. Take a p-Sylow subgroup S_p such that $S_p \vee S_q$ is not nilpotent. S_q is not self-conjugate in $S_p \vee S_q$. Let D be the maximal normal subgroup of $S_p \vee S_q$, which is contained in S_q. Take a subgroup U of S_q, covering D. Then $S_p \vee D$ is an LM-group and hence an M-group as before. Hence S_p is an abelian P-group. If $S_q = \{b\}$,

$$b\,a\,b^{-1} = a^\alpha,\quad \alpha \not\equiv 1,\quad \alpha^{q^\beta} \equiv 1 \pmod{p}$$

for any element $a \in S_p$. Suppose now that $S_p \vee S_q$ is not nilpotent and $p > q$. If we take an r-Sylow subgroup S_r ($r \neq q, p$), $S_p \vee S_r$ is nilpotent. Suppose to the contrary that $S_p \vee S_r$ is not nilpotent. If $p > r$, S_r would be cyclic. We may assume without loss of generality that $q < r$. Then $(a^{-1}\,b\,a)\,b^{-1} = a^{\alpha-1}$ would be contained in the normalizer of S_r. Since $\alpha \not\equiv 1 \pmod{p}$, this would imply that $S_p \vee S_r$ is nilpotent. If $p < r$, S_p would be cyclic, and for any element c of S_r we would have

$$a\,c\,a^{-1} = c^\gamma,\ b\,c\,b^{-1} = c^\beta,\ b\,a\,b^{-1} = a^\alpha,\ \gamma \not\equiv 1 \pmod{r}.$$

Hence

$$c^{\beta\gamma} = b\,c^\gamma\,b^{-1} = b\,(a\,c\,a^{-1})\,b^{-1} = a^\alpha\,c^\beta\,a^{-\alpha} = c^{\beta\gamma^\alpha}$$

or

$$\gamma^\alpha \equiv 1 \pmod{r}.$$

Hence $\alpha - 1 \equiv 0 \pmod{p}$, which contradicts the assumption $\alpha \not\equiv 1 \pmod{p}$.

Hence a directly indecomposable UM-group is either a modular p-group or a group of the type described in this theorem. In the latter case we have verified the properties (1) to (3). If $\beta_i = \beta_j$ for some $i \neq j$, $S_i \cup S_j \cup Q$ would not be an UM-group. Hence we have proved the first half of this theorem. It is not difficult to prove the converse if we observe that two q-subgroups Q_1 and Q_2 are transformed into each other by an element of prime order when $Q_1 \cap Q_2$ is maximal in both Q_1 and Q_2.

Using the same method as in the case of modular groups, SATO [2] determined the structure of locally finite UM-groups as follows.

Theorem 21. *A locally finite group G is an UM-group if and only if G is a direct product of modular p-groups and groups with the structure given in theorem* 20 *such that for each prime p elements of order p cannot appear in two distinct direct factors. If a direct factor is not a p-group, then in the notation of theorem* 20, *we may have an infinite abelian group P_i, but the number r is finite.*

The last assertion is a consequence of the property (4).

The structure of UM-groups containing an element of infinite order is not yet known. SATO [2] proved that *an UM-group G is an M-group if G contains an element of infinite order, all subgroups, $\neq e$, of G are not perfect, and every factor group of any subgroup of G is finite whenever this factor group is a J-group.*

7. Complemented groups.

A group is called a *K-group* if the subgroup lattice is complemented. Even the structure of finite K-groups is not yet known. A finite K-group is not necessarily solvable, and subgroups of a K-group are not always K-groups. Examples proving these statements are supplied by the alternating group of five letters and by the symmetric group of four letters. We have the following general properties.

Proposition 1.14. *The Φ-subgroup of a finite K-group consists of the identity element only.*

This is an immediate consequence of the so-called basis theorem (Cf. ZASSENHAUS [1] p. 46).

Proposition 1.15. *A nilpotent group is a K-group if and only if it is elementary abelian.*

This proposition follows from (1.14). Now consider the maximal nilpotent normal subgroup L of a finite K-group G. For any maximal subgroup M of G, $L \cap M$ is a normal subgroup of G and $L/L \cap M$ is isomorphic with one of the factor groups of a principal series. Since $\Phi(G) = e$, L must be a direct product of minimal abelian normal subgroups of G. By assumption G is a K-group, so that L has a complement

$$H: G = L \cup H \text{ and } L \cap H = e.$$

We can prove that H is a K-group. Since $G/L \cong H$, it is necessary to show that G/L is a K-group. Take a subgroup U of G which contains L. By assumption, U has a complement V in $L(G)$, namely V satisfies the relations $U \cup V = G$, and $U \cap V = e$. By the modular law, we have

$$U \cap (V \cup L) = (U \cap V) \cup L = L.$$

Hence $V \cup L/L$ is a complement of U/L in $L(G/L)$. We have now the following theorem (ZACHER [5]).

Theorem 22. *A finite group G is a K-group if and only if the maximal nilpotent normal subgroup L of G is a direct product of minimal normal abelian subgroups of G and L has a complement which is a K-group.*

Proof. The necessity of these conditions has been proved. Suppose now that a finite group G satisfies the above conditions. Let H be a complement of L. Then $G/L \cong H$ and by assumption G/L is a K-group. Take a subgroup U of G. Since G/L is a K-group, there exists a subgroup V of G such that $(U \cup L) \cup V = G$ and $(U \cup L) \cap V = L$. By the modular law, we have $(U \cup L) \cap V = (U \cap V) \cup L$. Hence $U \cap V \subseteq L$, or $U \cap V = U \cap L$. Take a normal subgroup N of G such that $L \supseteq N$, $N \cap U = e$ and suppose N is maximal under these conditions. We shall prove that $W = N \cup (V \cap H)$ is a complement of U. First of all

$$\begin{aligned} U \cap W &= U \cap \bigl(N \cup (V \cap H)\bigr) = U \cap \bigl((N \cup H) \cap V\bigr) \\ &= U \cap L \cap (N \cup H) = U \cap \bigl(N \cup (L \cap H)\bigr) = U \cap N = e. \end{aligned}$$

Consider now $U \cup W$. If $U \cup W \supseteq L$, we have

$$U \cup W = U \cup W \cup L = U \cup L \cup (V \cap H) = U \cup L \cup V = G.$$

If $U \cup W \not\supseteq L$, we could take a maximal subgroup M containing $U \cup W$. Since $U \cup W \cup L = G$, M would not contain L. Hence there would be a minimal normal subgroup T of G such that

$$T \subseteq L, \ T \cup M = G \text{ and } T \cap M = e,$$

because L is a direct product of minimal normal subgroups. $N \cup T$ would be a normal subgroup of G and would satisfy the relations $N \cup T \subseteq L$ and $(N \cup T) \cap U = e$. This contradicts the maximality of N. Hence $U \cup W = G$ and W is a complement of U.

This theorem tells us nothing about the structure of semi-simple K-groups. For a solvable group, we prove however the following theorem (ZACHER [5]).

Theorem 23. *A finite solvable group G is a K-group, if and only if G contains a series of normal subgroups*

$$e = N_0 \subset N_1 \subset \cdots \subset N_r = G$$

such that each N_{i+1}/N_i is a maximal nilpotent normal subgroup of G/N_i, and $\Phi\ (G/N_i) = e$ for $i = 0, 1, \ldots, r-1$.

Proof. Suppose G is a K-group. Then $\Phi(G) = e$ and the factor group G/N_1 is again a K-group. Hence our conditions are necessary.

Suppose conversely that G possesses a series of normal subgroups, satisfying the conditions of this theorem. Using induction on r, we have only to show that N_1 has a complement in G. Since $\Phi(G) = e$, N_1 is a direct product of minimal abelian subgroups. Let L be a maximal subgroup of G such that $L \subsetneq N_1$ and L has a complement H in G. We want to show that $L = N_1$. By way of contradiction, suppose $N_1 \neq L$. $N_1 \cap H$ contains a minimal normal subgroup K. K has a complement M. By the modular law we have

$$(K \cup L) \cap M \cap H = \big(K \cup (L \cap H)\big) \cap M = K \cap M = e,$$

and

$$(K \cup L) \cup (M \cap H) = L \cup \big((K \cup M) \cap H\big) = L \cup H = G.$$

Hence $M \cap H$ is a complement of $K \cup L$. This contradicts the maximality of L. Hence $L = N_1$ and N_1 has a complement H.

The structure of *finite super-solvable K-groups* has been determined by P. HALL [1].

Theorem 24. *The following properties of the finite group G are equivalent:*

(1) *G is a super-solvable K-group;*

(2) *G is isomorphic with a subgroup of a direct product of groups with square-free order,*

(3) *every subgroup U of G has a complement V such that $U V = V U$.*

Let us assume that (1) is satisfied. We want to prove that G is isomorphic with a subgroup of a direct product of groups with square-free order. If G contains two normal subgroups M and N such that $M \cap N = e$, then G is isomorphic with a subgroup of the direct product of G/M and G/N. Hence it is sufficient to prove that the order of G is square-free if G contains only one minimal normal subgroup. By assumption G is super-solvable, and so the p-SYLOW subgroup S is normal where p is the largest prime divisor of the order of G. By theorem 22, the maximal normal nilpotent subgroup is a direct product of minimal abelian normal subgroups. In our case G contains only one minimal subgroup, which is of order p. Hence S must be a minimal normal subgroup and so of order p. By a theorem of P. HALL, G contains a subgroup H such that $G = S\,H$ and $S \cap H = e$. Now the centralizer Z of S is a normal subgroup of G and $Z = S \times (H \cap Z)$. We have $H \cap Z = e$, as $H \cap Z$ is a normal subgroup of G contained in H. This implies that H is cyclic as it is isomorphic with a subgroup of the group of all automorphisms of a cyclic group S. On the other hand, H is a K-group by theorem 22. Hence all SYLOW subgroups of H are elementary abelian, which implies that the order of G is square-free. This proves (2).

Next assume the validity of (2). First of all we remark that a group with square-free order satisfies the condition (3). This is a consequence of HALL's extension of SYLOW's theorem. Let G be a direct product of two groups H and K both of which satisfy the condition (3). We shall show that G itself satisfies (3). Take a subgroup U of G. Every element u of U is of the form $h\,k$ ($h \in H$ and $k \in K$). The totality of elements of H which appear as H-factors of elements of U forms a subgroup V of H. Put $W = U \cap K$. By assumption V and W have complements V' and W' such that

$$H = V\,V', \quad V \cap V' = e, \quad W\,W' = K \text{ and } W \cap W' = e.$$

Now we prove that $V' \times W'$ is a complement of U in G. Consider an element u of $U \cap (V' \times W')$. u is a product $h\,k$ of $h \in H$ and $k \in K$. Since $u \in U$, h is in V and at the same time in V'. Hence $h = 1$ and $u = k$ is contained in $W = U \cap K$. Since $k \in W'$, we must have $u = k = 1$. This shows that

$$U \cap (V' \times W') = e.$$

Let $g = h\,k$ be an element of G ($h \in H$, $k \in K$). There are elements $v \in V$ and $v' \in V'$ such that $h = v\,v'$. By the definition of V, there is an element u of U such that $u = v\,k_1$ ($k_1 \in K$). Suppose $k_1^{-1}\,k = w\,w'$, where $w \in W$ and $w' \in W'$. Then

$$g = h\,k = (v\,k_1)\,k_1^{-1}\,v'\,k = (v\,k_1\,w)\,v'\,w' \in U(V' \times W').$$

Hence $G = U\,(V' \times W')$. Suppose next that a group G satisfies (3). Take subgroups U and V of G such that $U \supseteq V$. If we take a complement V' of V such that $G = V\,V'$ and $V \cap V' = e$, we have $U = V\,(U \cap V')$ and $V \cap (U \cap V') = e$. This shows that every subgroup of G satisfies (3), if G does. Hence the condition (2) implies (3).

Assume finally that G satisfies the condition (3). Applying a theorem of HALL we know that G is solvable, since G has SYLOW complements for all prime divisors of its order. We shall prove the condition (1). Let

$$G = N_0 \supset N_1 \supset \cdots \supset N_r = e$$

be a principal series of G. As G is solvable, each factor group N_{i-1}/N_i is abelian. Consider a subgroup H such that $N_{i-1} \supseteq H \supseteq N_i$. By assumption G contains a complement K such that $H K = G$ and $H \cap K = e$. The subgroup U defined by $U = (N_i \cup K) \cap N_{i-1}$ is a normal subgroup of G, and $N_{i-1} \supseteq U \supseteq N_i$. Hence $U = N_{i-1}$ or N_i. If $U = N_{i-1}$, we have

$$H = H \cap U = (N_i \cup K) \cap H = N_i \cup (K \cap H) = N_i.$$

If $U = N_i$, we have

$$H = H \cup U = H \cup N_i \cup (K \cap N_{i-1}) = H \cup (K \cap N_{i-1})$$
$$= (H \cup K) \cap N_{i-1} = N_{i-1}.$$

In any case each N_{i-1}/N_i contains no proper subgroup and hence it is a cyclic group of prime order. Thus we have shown that G is super-solvable.

For infinite groups satisfying the condition (3) of theorem 24, see BAEVA [*1*].

A group is called an $R\,K$-group if its subgroup lattice is relatively complemented. An $R\,K$-group G has the following property: Let

$$G \supseteq U \supseteq V \supseteq W$$

be a series of subgroups such that V is a normal subgroup of U. Then W is a normal subgroup of U whenever W is normal in V. This property may be verified as follows. By assumption there exists a relative complement K of V in the interval U/W. The isomorphism theorem asserts that W is a normal subgroup of K. Hence W is normal in U if W is normal in V. A group satisfying the above condition is called a *t-group*. A t-group of finite order is always super-solvable. Using this concept we formulate

Theorem 25. *A finite group G is an $R\,K$-group, if and only if G is a t-group with elementary abelian* SYLOW *subgroups* (ZACHER [*4*]).

Proof. If G is an $R\,K$-group, then it follows from (1.15) that SYLOW subgroups are elementary abelian. G is also a t-group as proved above.

Conversely assume that G is a t-group with elementary abelian SYLOW subgroups. Then G is solvable, and every nilpotent subgroup of G is elementary abelian. Hence it follows from theorem 23 that G is a K-group. To prove our assertion it suffices to show that any interval G/H is complemented. Let U be any subgroup containing H. We shall prove the existence of a complement of U in G/H by using induction on the order g of G. If the order of H is divisible by the largest prime divisor p of g, then H contains a subgroup P of order p. P is by assumption a normal subgroup of G, since the p-SYLOW subgroup S of G is normal. Hence we may apply inductive hypothesis on G/P to prove the existence of a complement of U. So assume that the order of U is prime to p. Since G is solvable, U contains a p-SYLOW complement V containing H, and G has a p-SYLOW complement W which contains V. Being isomorphic with G/S, W is an R K-group by inductive hypothesis. Hence there is a subgroup L such that $V \cup L = W$ and $V \cap L = H$. S is elementary abelian, so that there is a complement T of $S \cap U$ in S. T is a normal subgroup of G since T is a normal subgroup of S. Letting $K = T\,L$, we shall show that K is a relative complement of U. First of all we have

$$K \cup U = T \cup L \cup V \cup (U \cap S) = T \cup (U \cap S) \cup L \cup V$$
$$= S \cup W = G.$$

Consider an element a in $U \cap K$. Then $a = v\,s$ where $v \in V$ and $s \in S \cap U$. On the other hand, $a \in K$ implies that $a = l\,t$ where $l \in L$ and $t \in T$. Hence $a = v\,s = l\,t$ or $v^{-1}\,l = s\,t^{-1}$. Since $v^{-1}\,l \in W$ and $s\,t^{-1} \in S$, we have $v^{-1}\,l = s\,t^{-1} = 1$, i. e. $v = l$ and $s = t$. Since $(S \cap U) \cap T = e$, we must have $s = t = 1$ and $a \in V \cap L = H$. Hence $K \cap U \subseteq H$. On the other hand $H \subseteq K \cap U$ by definition, so that $U \cap K = H$. This shows that K is a relative complement of U.

As a corollary we obtain

Corollary. *A finite group G is an R K-group if and only if G is a solvable K-group satisfying the following condition:*

A chain of subgroups

$$G = U_0 \supseteq U_1 \supseteq \cdots \supseteq U_r = e$$

is a principal series of G whenever each $[U_{i-1}:U_i] = p_i$ is a prime and $p_1 \leq p_2 \leq \cdots \leq p_r$.

Chapter II.

Isomorphisms of subgroup lattices.

1. Projectivities.

An isomorphic mapping of the subgroup lattice $L(G)$ of a group G onto the subgroup lattice $L(H)$ of a group H is called a *projectivity* of G onto H. (Several different terms have been used instead of projectivity. For instance, ORE and SADOVSKY called it a *structural isomorphism*, and JONES and SUZUKI used the term *L-isomorphism*.) Every isomorphism of the group G onto H induces clearly an isomorphism of the lattice $L(G)$ onto $L(H)$. The converse assertion is not always true, since there exist large classes of non-isomorphic groups whose subgroup lattices are isomorphic. Under what circumstances, then, is a projectivity induced by a group isomorphism? The following theorem reduces this problem largely to the corresponding problem for finitely generated groups.

Theorem 1. *Let φ be a projectivity of a group G, and let K be a collection of subgroups G_α such that*

(1) *each G_α has a finite set of generators,*

(2) *for G_α, $G_\beta \in K$, there exists a G_γ in K such that $G_\alpha \cup G_\beta \subseteq G_\gamma$, and*

(3) *every element of G is contained in some $G_\alpha \in K$.*

If φ is induced by a group isomorphism on each $G_\alpha \in K$, then φ is induced by a group isomorphism of G (SADOVSKY [*1*], theorem 1).

Proof. For any group H, the number of isomorphisms of H which induce a given projectivity is determined by the number of automorphisms of H, which induce the identical auto-projectivity of H. Hence in particular if H is finitely generated, the number of such automorphisms is finite.

Let A_α be the totality of isomorphisms of G_α which induce φ. We shall introduce a partial order in $S = \{A_\alpha\}$ by defining $A_\alpha \subseteq A_\beta$ if and only if $G_\alpha \subseteq G_\beta$. By assumption S is then a directed set. If $A_\alpha \subseteq A_\beta$, we define a projection $\Pi_{\alpha\beta}$ in the following way. Take a $\sigma_\beta \in A_\beta$, then σ_β induces φ on G_β, so that the contraction of σ_β on G_α induces φ. Hence it is an element σ_α of A_α. We define $\Pi_{\alpha\beta}(\sigma_\beta) = \sigma_\alpha$. Clearly this mapping satisfies the following conditions:

(1) $\Pi_{\alpha\alpha}$ is the identical mapping of A_α,

(2) if $A_\alpha \subseteq A_\beta$ and $A_\beta \subseteq A_\gamma$, then $A_\alpha \subseteq A_\gamma$ and $\Pi_{\alpha\gamma} = \Pi_{\alpha\beta}\,\Pi_{\beta\gamma}$, and

(3) for any α and β, there is a γ such that $\Pi_{\alpha\gamma}$ and $\Pi_{\alpha\gamma}$ exist.

By a theorem of KUROSCH-STEENROD (Amer. J. Math. **58**, p. 661) the inverse limit is not empty, since spaces with only finitely many elements are compact, and lim A_α is the set of isomorphisms of G which induce φ. Thus this theorem is proved.

2. Projectivities of abelian groups.

Projectivities of abelian groups were studied by BAER [*6*].

Proposition 2.1. *Let G be an abelian p-group. If there exists a projectivity of G onto a group H, which is not a p-group, then G is an elemen-*

tary abelian group and H is a generalized P-group (BAER [6], p. 23, theorem 11.2).

Proof. Since G is abelian, H is a locally finite M-group. Hence the results of I. § 5 can be applied, and this proves our proposition.

Sometimes projectivities are induced by a mapping of elements, which need not be an isomorphism, but has a similar property (BAER [6, 8] and SADOVSKY [1, 2, 3]).

Proposition 2.2. *Let G be a direct product of two cyclic groups X and Y, and let φ be a projectivity of G onto a group H. If x' and y' are generators of $\varphi(X)$ and $\varphi(Y)$ respectively, then there exist generators x and y of X and Y respectively such that φ maps the cyclic group $\{x\, y\}$ onto $\{x'\, y'\}$.*

Proof. Take the subgroup Z of G such that $\varphi(Z) = \{x'\, y'\}$. Let the generator of Z be $x\, y$ $(x \in X,\ y \in Y)$. Since

$$x' \in \{x'\, y', y'\} = \{x'\, y'\} \cup \{y'\},$$

we have $X \subseteq Z \cup Y$, or $X \subseteq \{x\} \cup Y$. Hence $\{x\} = X$ and similarly $\{y\} = Y$.

A similar theorem is valid in the case of a free product (SADOVSKY [3]). *If $G = \{a\} * \{b\}$ is a free product, and if φ is a projectivity of G onto another group G', there exist generators a', b' of $\varphi\{a\}$ and $\varphi\{b\}$ respectively, such that $\varphi\{a\, b\} = \{a'\, b'\}$.*

Proposition 2.3. *Let G be an abelian group, and let φ be a projectivity of G onto a group H. Assume that G contains two elements u and x such that $\{u\} \cap \{x\} = \{u\, x\} \cap \{x\} = e$. Take a generator u' of $\varphi\{u\}$. If the order of u is either p^n $(n > 1)$ or infinite, there is a unique element $x' = f(x, u, u')$ of H such that $\varphi\{x\} = \{x'\}$ and $\varphi\{u\, x\} = \{u'\, x'\}$.*

Proof. The uniqueness is rather trivial. Suppose we get two solutions x' and x''. Then $u'\, x' = (u'\, x'')^k$ for some k, and k may be chosen positive. Hence

$$u'\, x' = u'\,(x''\, u')^{k-1}\, x^{-1} x'',$$

or

$$x'\, x''^{-1} = (x''\, u')^{k-1} = (u'\, x'')^{k-1} \in \varphi\{x\} \cap \varphi(u\, x\} = e.$$

We shall now prove the existence of x'. Take an element v such that $\{v\} = \varphi\{x\}$. Then the number of subgroups of the form $\{u'\, v\}$ is equal to the number of generators of $\{v\}$. If x is of finite order, or if u is of finite order, $\{x\}$ has the same number of generators as $\{v\}$. Hence by (2.2) there is at least one solution x'. If the order of u is infinite but that of x is finite, $\varphi\{x\}$ is normal in $\varphi\{u, x\}$. Hence $\varphi\{u\, x\} = \{u'^{\,i}\, y\}$ where $\{y\} = \varphi\{x\}$ and $\{u'^{\,i}\} = \{u'\}$. Since $\{u\}$ is infinite cyclic, we may take $i = 1$.

Proposition 2.4. *Let $G = \{u\} \times S$ be an abelian group. Suppose that the assumptions of* (2.3) *are satisfied for u and $s \in S$. Then the mapping f gives a one-to-one mapping of S onto $\varphi(S)$.*

Proof. Let $y \in \varphi(S)$. Then there exist elements z and v in G such that

$$\varphi\{z\} = \{y\},\ \varphi\{v\} = \{u'\} \text{ and } \varphi\{v\,z\} = \{u'\,y\}.$$

$\{u\} = \{v\}$ implies $v^k = u$ and $\{z^k\} = \{z\}$. Since G is abelian, $(v\,z)^k = u\,z^k$ generates $\{v\,z\}$. Hence $y = f(z^k; u, u')$, and f is onto.

Proposition 2.5. *Under the same assumption as in* (2.4), *we assume moreover that, for any $s \in S$, there exists a $t \in S$ such that*

$$\{s\} \cap \{t\} = \{s\,t\} \cap \{s\} = e.$$

If $\varphi(G)$ is abelian, f is an isomorphism of S onto $\varphi(S)$.

We prove here this proposition under the further assumption that if the order of u is infinite, S contains an element of infinite order.

Proof. Take an element $s \in S$. By assumption there is a $t \in S$ such that $\{s\} \cap \{t\} = \{s\,t\} \cap \{s\} = e$. Hence by (2.3) there exists a unique

$$s' = f(s;\ t, t'), \text{ where } t' = f(t;\ u, u').$$

In a similar way as in the proof of (2.3), we can prove that

$$f(t\,s;\ u, u') = f(t;\ u, u')\, f(s;\ t, t')$$

and

$$f(s;\ t, t') = f(s;\ u\,t, u'\,t').$$

Now $f(s; u\,t, u'\,t') = f(s; u, u')$ by the commutativity of $\varphi(S)$. Putting $f(s; u, u') = f(s)$, we have thus proved that

$$\text{if } \{s\} \cap \{t\} = \{s\,t\} \cap \{s\} = e, \text{ then } f(t\,s) = f(t)\, f(s).$$

Hence for powers of s,

$$f(t\,s^i) = f(t\,s\,s^{i-1}) = f(t\,s)\, f(s^{i-1}) = f(t)\, f(s)\, f(s^{i-1}),$$

or by induction

$$f(s^i) = f(s)^i.$$

If $\{s\} \cap \{t\} = e$ and if one of them has infinite order, we have $f(s\,t) = f(s)\, f(t)$. The same is true if both orders are powers of the same prime. If u is of infinite order, S contains an element v of infinite order by assumption. Then

$$f(v\,s\,t) = f(v)\, f(s\,t) = f(v\,s)\, f(t) = \big(f(v)\, f(s)\big)\, f(t),$$

or

$$f(s\,t) = f(s)\, f(t) \text{ for every } s \text{ and } t.$$

If u is of prime power order, S is also a p-group, and $\{s, t\}$ is a direct product of at most two cyclic groups generated by x and y respectively. Let $s = x^i y^j$ and $t = x^k y^l$. Then

$$f(s\,t) = f(x^{i+k}\,y^{j+l}) = f(x)^{i+k}\,f(y)^{j+l} = f(s)\,f(t).$$

Hence we have proved (2.5) in all cases.

Theorem 2. *Let G be an abelian p-group which satisfies the following conditions: if G contains an element of order p^n, G contains at least three independent elements of such order. Then any projectivity of G onto another abelian group H is induced by an isomorphism* (BAER [*6*], p. 27, Lemma 11.5).

Proof. Let K be the totality of subgroups G_α of G, such that each G_α is finitely generated and if p^n is the largest order of elements of G_α, G_α contains at least three independent elements of order p^n. Then $\bigcup_\alpha G_\alpha = G$ and K forms a directed subset of $L(G)$. By theorem 1 of p. 32, it is sufficient to prove our theorem in each G_α. Hence we may assume that G itself is finitely generated, and hence a *finite* group.

Let p^n be the largest order of elements of G. G is a direct product of cyclic subgroups and at least three direct components are of order p^n. Take one of them, say $u: G = \{u\} \times S$. Then the assumptions of (2.5) are satisfied. Hence for any projectivity φ, $S \cong \varphi(S)$ and this isomorphism induces φ. Put $\varphi\{u\} = \{u'\}$. We want to show that the mapping of G, defined by $u \to u'$ and $s \to s' = f(s; u, u')$ $(s \in S)$, induces φ. Take another basis element v of order p^n. Then $S = \{v\} \times B$. It is sufficient to prove that $f(u; v, v') = u'$. Now

$$\varphi\{u\,v\} = \{u'\,f(v;\ u, u')\} = \{v'\,f(u;\ v, v')\}.$$

By uniqueness of f, we have $f(u; v, v') = u'$.

In this case the number of isomorphisms of G which induce φ, is equal to the number of automorphisms of G such that $U^\sigma \subseteq U$ for all subgroups of G. These automorphisms have been determined by BAER (Amer. J. of Math. **59**(1937)). Cf. also F. W. LEVI (J. Indian Math. Soc. **10** (1946)).

Theorem 3. *Let G be an abelian group, which contains at least two elements u and v of infinite order, such that $\{u\} \cap \{v\} = e$. Then any projectivity of G is induced by exactly two isomorphisms* (BAER [*6*], p. 30, theorem 12.1).

Proof. Let φ be a projectivity of G onto H. Then since H is an M-group, H is abelian. As before, we may assume that G is finitely generated. Take one element u of infinite order, such that $G = \{u\} \times S$. The mapping $f(u; u, u')$, defined in (2.3), gives an isomorphic mapping of S onto $\varphi(S)$. Moreover $f(s; u, u') = f(s; u, u'^{i})$. Hence the isomorphism of G, defined by $u \to u'$ and $s \to s'$, induces φ. This shows the existence of an isomorphism which induces φ. It is almost obvious that there are exactly two such isomorphisms.

The projectivities of finite M-groups have been completely investigated by JONES [1]. According to his results (see section 5 below, in particular theorem 7, p. 39), the study of the subgroup lattice of an M-group is reduced, essentially, to the study of the abelian case. In the subgroup lattices of abelian p-groups, any interval containing a proper neutral element is a chain. A modular lattice satisfying this property is said to be *primary*. INABA [1, 2] considered such modular lattices. If a primary lattice L is of finite dimension, then the greatest element is a union of maximal chains, independent of each other in the usual sense. Let m be the number of chains with the largest dimension, which appear in an expression of $I = a_1 \cup \cdots \cup a_m$. Then it is proved that if $m \geq 4$, *L is isomorphic with the lattice of all submodules in a submodule M' of an R-module M, where R is a uniserial, completely primary ring.* The subgroup lattices of finite abelian p-groups are also studied by BAER [7] and RIBEIRO [1]. GRAYEV [1] studied the projectivities of separable locally compact abelian groups, replacing $L(G)$ by the lattice formed by all closed subgroups.

3. Projectivities of locally free groups.

Theorem 4. *Every locally free group is determined by its subgroup lattice. If a locally free group is not locally cyclic, every projectivity is induced by exactly one isomorphism* (SADOVSKY [1], theorem 2).

We sketch a proof of this theorem. By theorem 1, it is sufficient to prove this theorem for a free group with a finite set of generators, and we may assume, moreover, that it is not cyclic. Let F_2 be a free group generated by two elements; $F_2 = \{a\} * \{b\}$, and let φ be a projectivity of F_2. We can take suitable generators a and b of F_2 such that $\varphi\{a\} = \{a'\}$, $\varphi\{b\} = \{b'\}$ and $\varphi\{a\,b\} = \{a'\,b'\}$. This can be proved by using, for instance, a theorem of MAGNUS. If we have chosen these suitable generators, it can be proved that φ maps $\{a^{\varepsilon_1} b^{\varepsilon_2}\}$ onto $\{a'^{\varepsilon_1} b'^{\varepsilon_2}\}$ ($\varepsilon_i = \pm 1$). Using these lemmas, we can prove that if a projectivity φ of F_4, a free group with four generators a, b, c, and d, maps $\{a\}$, $\{b\}$, $\{c\}$ and $\{d\}$ onto $\{a'\}$, $\{b'\}$, $\{c'\}$ and $\{d'\}$ respectively, and if moreover $\varphi\{a\,b\} = \{a'\,b'\}$, then for a suitable choice of c and d, we have $\varphi\{a\,c\} = \{a'\,c'\}$, $\varphi\{ad\} = \{a'\,d'\}$ and $\varphi\{c\,d\} = \{c'\,d'\}$. Let G be a free group with n generators. Suppose, taking suitable generators $a, b, \ldots$ of G, that $\varphi\{a\} = \{a'\}$, $\varphi\{b\} = \{b'\}, \ldots$, and $\varphi\{a\,b\} = \{a'\,b'\} \cdots$. Then we can prove that $\varphi\{a^{\alpha} b\} = \{a'^{\alpha} b'\}$ by induction on α. Hence $\varphi\{a^{\alpha}\} = \{a'^{\alpha}\}$. Now $a', b', \ldots$ are the generators of $G' = \varphi(G)$, so that every element g' of G' can be expressed as a product of powers of these elements $a', b', \ldots$. We take one of the shortest expressions of g', say $g' = a'^{\alpha} b'^{\beta} \cdots c'^{\gamma}$. Then φ maps $\{g\}$ onto $\{g'\}$, where $g = a^{\alpha} b^{\beta} \cdots c^{\gamma}$. This may be proved by induction on the length of the expression of g'. Let $g' = a' h'$. Then $\varphi\{h\} = \{h'\}$ where $h = b^{\beta} \cdots c^{\gamma}$. Since there is no relation between a^{α} and g, we have $\varphi\{a^{-\alpha\theta} g^{\varepsilon}\} = \{h'\}$ or $g^{\varepsilon} = a^{\alpha\theta} (q^{\beta} \cdots c^{\gamma})^{\nu}$ where $\varepsilon, \theta, \nu = \pm 1$. Similarly if $g' = k' c'^{\gamma}$, we have $g^{\varepsilon_1} = (a^{\alpha} \cdots)^{\nu_1} c^{\gamma\theta_1}$, where $\varepsilon_1, \theta_1, \nu_1 = \pm 1$. Since G is a free group, we conclude that $g = a^{\alpha} b^{\beta} \cdots c^{\gamma}$.

It can be seen that the mapping defined by $g \to g'$ is an isomorphism of G onto G'. Hence we have theorem 4.

Sadovsky [*2*, *3*] extended these results to the case of free products. Let $G = A * B$ be a (proper) free product of two groups $A \neq 1$ and $B \neq 1$, and let φ be a projectivity of G. Then $\varphi(A) \cong A$, $\varphi(B) \cong B$ and $\varphi(G) = \varphi(A) * \varphi(B)$, so that *every proper free product is determined by its subgroup lattice*. It is, however, not yet known whether φ is induced by an isomorphism or not.

4. Projectivities of finite groups.

Jones [*1*] gave a necessary and sufficient condition for two finite groups to have the same lattice as their subgroup lattices.

Following Jones we call a set $(a_1, \ldots, a_n)$ of elements of a finite group G a *power basis* of G if every subgroup $\{a_i\}$ is a maximal cyclic subgroup of G, $\{a_i\} \neq (a_j)$ $(i \neq j)$ and if G has no other maximal cyclic subgroup. Let $(a_1, \ldots, a_n)$ be a power basis of G. The set $(0_1, \ldots, 0_n)$ of integers $0_1, \ldots, 0_n$ is called *the set of power basis orders of* A, if each 0_i is the order $0(a_i)$ of a_i. This set is clearly an invariant of G.

An *F-system* is a set of positive integers $p_{ij}, q_{ij}, \varepsilon_{ij}$ $(i = 1, 2, \ldots, n;$ $j = 1, 2, \ldots, n_i)$ such that $p_{i1}, \ldots, p_{in_i}$ are different primes for every i, and similarly $q_{i1}, \ldots, q_{in_i}$ are primes and different to each other. Put $P_i = \prod_j p_{ij}^{\varepsilon_{ij}}$ and $Q_i = \prod_j q_{ij}^{\varepsilon_{ij}}$.

Let F be an F-system. We shall call the triple of integers (x, y, i) *F-allowable*, if there do not exist any integers β_{ij}, γ_{ij} with $\beta_{ij} \neq \gamma_{ij}$ for at least one pair (i, j) and such that g. c. d. $(x, P_i) = \prod p_{ij}^{\beta_{ij}}$ and g. c. d. $(y, Q_i) = \prod q_{ij}^{\gamma_{ij}}$.

Using these definitions Jones' criterion (Jones [*1*], p. 543, theorem 2.1) reads as follows.

Theorem 5. *Two finite groups A and B have isomorphic subgroup lattices, if and only if there exists an F-system $F: p_{ij}, q_{ij}, \varepsilon_{ij}$ such that*

(1) *P_i and Q_i form the set of power basis orders for A and B respectively,*

(2) *if we take power basis* $\{a_i\}$ *and* $\{b_i\}$ *of* A *and* B *respectively such that* $0(a_i) = P_i$ *and* $0(b_i) = Q_i$, *then, for three* F*-allowable triples* (x, u, i), (y, v, j) *and* (z, w, k), $a_i^x a_j^y = a_k^z$ *implies* $b_k^w \in \{b_i^u, b_i^v\}$ *and similarly*

(3) $b_i^u b_j^v = b_k^w$ *implies* $a_k^z \in \{a_i^x, a_j^y\}$.

Proof. Assume that there is a projectivity φ of A onto B. Take a power basis $(a_1, \ldots, a_n)$ of A. Then by theorem I. 2, $\varphi\{a_i\}$ is also a maximal cyclic subgroup of B. Hence for a suitable ordering, we may have $\varphi\{a_i\} = \{b_i\}$, and $(b_1, \ldots, b_n)$ forms a power basis of B. Let $P_i = 0(a_i)$. Let $P_i = \prod p_{ij}^{e_{ij}}$ be the decomposition into a product of prime numbers. Then $\{a_i\}$ is a direct product of $\{a_i^{x_j}\}$ such that $0(a_i^{x_j}) = p_{ij}^{e_{ij}}$. Now $\varphi\{a_i\}$ is also a direct product of $\varphi\{a_i^{x_j}\}$ of order $q_{ij}^{e_{ij}}$. Then $0(b_i) = Q_i = \prod q_{ij}^{e_{ij}}$ and $F: p_{ij}, q_{ij}, \varepsilon_{ij}$ forms an F-system. It is now easily verified that φ maps $\{a_i^x\}$ onto $\{b_i^u\}$ if (x, u, i) is an F-allowable triple of integers. Since φ preserves unions, the necessity of our conditions is thus proved.

Suppose conversely that there exists an F-system satisfying conditions (1) to (3). Let E be the totality of sets $E(x_1, \ldots, x_n)$, where $E(x_1, \ldots, x_n)$ are the (set-theoretical) unions of elements of $\{a_1^{x_1}\}, \ldots, \{a_n^{x_n}\}$. Similarly we define the totality G of $G(u_1, \ldots, u_n)$, which are the unions of elements of $\{b_1^{u_1}\}, \ldots, \{b_n^{u_n}\}$. We define a mapping f of E onto G by setting $f E(x_1, \ldots, x_n) = G(u_1, \ldots, u_n)$, if triples (x_i, u_i, i) are F-allowable for all $i = 1, 2, \ldots, n$. E and G form partially ordered sets by usual inclusion relation. We want to show that f preserves the order. Suppose $f E(x_1, \ldots, x_n) = G(u_1, \ldots, u_n)$, $f E(y_1, \ldots, y_n) = G(v_1, \ldots, v_n)$ and $E(x_1, \ldots, x_n) \supseteq E(y_1, \ldots, y_n)$. Then for every k, there are two integers x and i such that $a_k^{y_k} = a_n^{x_i x}$. Let the triple (x, u, i) be F-allowable. Then by definition $(x_i x, u_i u, i)$ is F-allowable. Since $a_k^{y_k} = a_i^{x_i x} a_j^{P_j}$, $b_k^{v_k}$ is in the subgroup generated by $b_i^{u_i n}$ and $b_j^{Q_j}$, namely in $\{b_i^{u_i u}\}$. Hence $b_k^{v_k} \in \{b_i^{u_i u}\} \subseteq G(u_1, \ldots, u_n)$. Now every subgroup of A appears in E, and similarly the subgroup lattice $L(B)$ of B is contained in G. We shall show that f induces an isomorphism of $L(A)$ on $L(B)$. Assume $E(x_1, \ldots, x_n)$ to be a subgroup of A, and suppose $f E(x_1, \ldots, x_n) = G(u_1, \ldots, u_n)$. Take two elements $b_i^{u_i u}$ and $b_j^{u_j v}$ of $G(u_1, \ldots, u_n)$, and let

$$b_k^w = b_i^{u_i u} b_j^{u_j v}.$$

It follows from the condition (2) that

$$a_k^z \in \{a_i^{x_i x}, a_j^{x_j y}\}$$

for F-allowable triples $(x_i x, u_i u, i)$, $(x_j y, u_j v, j)$ and (z, w, k). Hence $a_k^z = a_k^{x_k t}$ by assumption. Now $x_k t \equiv z \pmod{P_k}$ implies $w \equiv u_k s \pmod{Q_k}$. This shows that $b_k^w \in \{b_k^{u_k}\}$ and hence $G(u_1, \ldots, u_n)$ forms a subgroup. The theorem is thus proved.

5. Projectivities of modular groups.

Projectivities of modular groups have been studied by BAER [*6, 10*], BEAUMONT [*1, 2*], JONES [*1*], SATO [*3*] and ZAPPA [*5*]. The fascinating feature of their studies is the existence of projectivities from a large class of M-groups upon abelian groups. We consider a locally finite M-group G. If G is not L-decomposable, then by theorem I.18 of p. 22, G is either a p-group or a P_0^*-group. If G is a P_0^*-group, G contains an abelian p-group P of exponent p as a normal subgroup, and the factor group G/P is a cyclic group of order q^n (q is a prime). In this case *G admits a projectivity upon an abelian group if and only if P is a maximal subgroup of G*; in other words, $n = 1$. If G is a Hamiltonian group, we can prove the following theorem (BAER [*6*], p. 11, theorem 6.1).

Theorem 6. *Let G be a primary Hamiltonian group. Every projectivity of G is induced by exactly four isomorphisms of G.*

Proof. By assumption G is a direct product of the quaternion group Q and an elementary abelian 2-group T. Consider any minimal subgroup U of T. The subgroup $Q \cup U$ contains exactly three minimal subgroups. This implies that for any projectivity φ of G, the image $\varphi(U)$ is of order 2 and

$$\varphi(Q \cup U) = \varphi(Q) \times \varphi(U).$$

Hence $\varphi(G) = \varphi(Q) \times \varphi(T)$ and $\varphi(G) \cong G$. It is not difficult to show that there exist exactly four isomorphisms which induce φ.

This theorem shows that a primary Hamiltonian group admits no projectivity upon an abelian group. The situation is, however, rather different in the case of a locally finite modular p-group which is not Hamiltonian. JONES [*1*] and SATO [*3*] have shown that the subgroup lattice of such a group is always isomorphic with that of a suitable abelian group. This result is, in a somewhat different form, obtained also by BAER [*10*]. If an M-group contains an element of infinite order, the situtaion is much simpler; in fact, SATO [*3*] has proved that *if an M-group G contains an element of infinite order, then G admits always a projectivity upon a suitable abelian group.* We shall prove here the existence of a projectivity, upon an abelian group, of a locally finite modular p-group which is not Hamiltonian, using the method employed by BAER [*10*].

Theorem 7. *Let G be a locally finite modular p-group. If G is not Hamiltonian, then G has a projectivity upon an abelian group H* (JONES [*1*], p. 313, theorem 3.13; SATO [*3*], theorem 1). *Moreover this projectivity is induced by a crossed isomorphism of H onto G* (BAER [*10*], p. 381, theorem 7.1).

Here a *crossed isomorphism* of an abelian group H onto G is a mapping τ which satisfies

$$\tau(a)\,\tau(b) = \tau(a^{f(b)}\,b)$$

where $f(x)$, for every x in H, is an automorphism of H such that

$$f(x)\,f(y) = f(x^{f(y)}\,y).$$

Proof. Let G be a locally finite modular p-group, which is not Hamiltonian. By theorem I. 18, G contains an abelian normal subgroup A_0 of bounded exponent with the property that for a suitable element z of G,

$$G = \{A_0, z\}$$

and $z^{-1}\,a\,z = a^{1+p^s}$ for any element a of A where s is independent of the choice of a and $s \geq 2$ in case $p = 2$. Denote by p^m the maximal order of elements of A_0. For any $g \in G$, and $a \in A_0$, we have $g^{-1}\,a\,g = a^{e(g)}$ where $g \to e(g)$ is a homomorphism of G into an additive group of integers modulo p^m. If we denote by A the kernel of this homomorphism, A is a normal subgroup of G containing A_0. Since A_0 is in the center of A and A/A_0 is cyclic, we conclude that A is abelian and the order of G/A is $p^{m-s} = p^n$. $z^{p^n} = u$ is now an element of A. Let

$$e = e(z) = 1 + p^s \text{ and } \sigma(k) = 1 + e + \cdots + e^{k-1}.$$

Then $\sigma(p^n)/p^n \equiv (e^{p^n} - 1)/(e-1)\,p^n \equiv e^* \pmod{p^m}$ and e^* is prime to p. Hence there exists a unique element v of A such that $v^{e^*} = u$, and there exists one, and essentially only one, abelian group B which is obtained by adjoining to A an element w, subject to the relation $w^{p^n} = v$.

Every element b of B may be expressed uniquely in the form $a\,w^i$ where $a \in A$ and $0 < i \leq p^n$. Let $i = i(b)$. Since the multiplicative order of e modulo p^m is p^n, it follows that $0 < j \leq p^n$ and $\sigma(j) \equiv 0 \pmod{p^n}$ imply $j = p^n$. Hence we conclude that if $0 < i, j \leq p^n$, then $i = j$ is a necessary and sufficient condition for $\sigma(i) \equiv \sigma(j) \pmod{p^n}$. Hence there exists to every element b of B a uniquely determined integer $j(b)$ such that

$$0 < j(x) \leq p^n \text{ and } i(b) \equiv \sigma\big(j(b)\big) \pmod{p^n}.$$

Now $b^{f(x)} = b^{e^{j(x)}}$ defines an automorphism of B. If $x = w$, $i(w) = 1$ and so $j(w) = 1$. Hence $f(w) = e$. If $x \in A$, we have $i(x) = p^n$ and $j(x) = p^n$. Hence $f(x) = 1$ for $x \in A$. Furthermore we have

$$f(x) = f(w^{i(x)}) = f(w^{i(x)+k p^n})$$

for any $x \in B$ and for every integer k. Hence we have for any $x, y \in B$, $f(x^{f(y)}\,y) = f(w^{i(x)\,f(y)+i(y)})$ and, since

$$i(x)\,f(y) + i(y) = \sigma(j(x))\,e^{j(y)} + \sigma(j(y)) = \sigma(j(x) + (j(y)),$$

we conclude that

$$f(x^{f(y)}\, y) = e^{j(x)+j(y)} = e^{j(x)}\, e^{j(y)} = f(x)\, f(y).$$

In B we introduce another multiplication by $x \circ y = x^{f(y)}\, y$. Then the above formula $f(x^{f(y)}\, y) = f(x)\, f(y)$ implies the associative relation

$$(x \circ y) \circ z = x \circ (y \circ z).$$

Since $f(x)$ is an automorphism of B, for any $x, y \in B$ we can find an element z such that $z^{f(x)} = y\, x^{-1}$. Hence z is a solution of $z \circ x = y$. Hence B forms a group under the new multiplication. We denote this new group by C. The relation $f(x^{f(y)}\, y) = f(x)\, f(y)$ implies that $f(x \circ y) = f(x)\, f(y)$, i. e. f defines a homomorphism of C into the group of automorphisms of B. The kernel of this homomorphism is A. C is thus generated by A and w. Denoting by w' the inverse element of w in C, we conclude that

$$w' \circ a \circ w = (w'\, a) \circ w = (w'\, a)^{f(w)}\, w = w'^{f(w)}\, a^{f(w)}\, w = w^{-1}\, a^{f(w)}\, w = a^{e}.$$

Furthermore we have

$$(w)^{p^n} = w^{\sigma(p^n)} = w^{p^n *} = v^{e*} = u.$$

Hence C is isomorphic with the original G under the mapping defined by $a \to a$ $(a \in A)$ and $w \to z$.

Thus we have proved that every locally finite modular p-group, which is not Hamiltonian, may be obtained from an abelian group by changing its multiplication suitably. The identical mapping τ from B to C in the above argument satisfies the identity $\tau(x)\, \tau(y) = \tau(x^{f(y)}\, y)$ where f is an automorphism of B satisfying $f(x^{f(y)}\, y) = f(x)\, f(y)$. Hence τ is a crossed isomorphism (Cf. Baer [*10*]). This proves the second part of this theorem.

If τ is a crossed isomorphism (of the type constructed above) of a group G onto a group H, and if S is a subgroup of G, then it is easy to show that $\tau(S)$ is a subgroup of H and vice versa. Hence τ induces an isomorphism of $L(G)$ onto $L(H)$.

In fact Baer [*10*] proved that in a fairly general case every projectivity of abelian groups is induced in this manner.

6. Index-preserving projectivities.

A projectivity φ of a group G is called *index-preserving* if

$$(U:V) = \big(\varphi(U):\varphi(V)\big),$$

for any *cyclic* subgroup U of G and every subgroup V of U. If this relation holds for *every* subgroup U of G, then φ is called *strictly index-preserving*. An index-preserving projectivity maps p-groups into p-groups, so that *every index-preserving projectivity of a finite group is strictly index-preserving*. It is not known whether there exist index-preserving projectivities which are not strictly index-preserving.

In the following we consider mainly *finite* groups.

Proposition 2.6. *Let φ be an index-preserving projectivity of a finite group G onto a group H. If N is a maximal normal subgroup of G, then $\varphi(N)$ is a normal subgroup of H.*

Proof. Consider first the case in which N is a maximal subgroup of G. We want to prove our assertion by using induction on the order of G. If G contains no maximal subgroup other than N, G is a cyclic group and the assertion follows from theorem I. 2 of p. 4. Assume next that G contains a maximal subgroup $M \neq N$. $D = M \cap N$ is a maximal subgroup of M and self-conjugate in M. Hence by inductive hypothesis $\varphi(D)$ is normal in $\varphi(M)$. If $\varphi(N)$ were not normal in H, some element of $\varphi(M)$ would transform $\varphi(N)$ into another subgroup N^*. We take a subgroup P of G such that $\varphi(P) = N^*$. Since $N^* \supseteq \varphi(D)$, we have $P \supseteq D$, or $P \cap N = D$. Hence D would be a normal subgroup of both P and M, and so of G. The factor group G/D would be a P-group of order $p\,q$, where p and q are primes, and $(N:D) = p$. By inductive hypothesis, $\varphi(D)$ would also be a normal subgroup of N^*, and so of H. φ induces now an index-preserving L-isomorphism of G/D onto $H/\varphi(D)$. Since G/D is a P-group, $\varphi(D)$ would be a normal subgroup of H, contrary to the assumption. Hence $\varphi(N)$ is self-conjugate.

In general, the factor group G/H is by assumption a simple group, so that there are subgroups U_λ of G such that N is a maximal subgroup of each U_λ and $\bigcup U_\lambda = G$. Since φ induces an index-preserving projectivity of each U_λ, $\varphi(N)$ is normal in $\varphi(U_\lambda)$ and from $\bigcup \varphi(U_;) = H$ it follows that $\varphi(N)$ is a normal subgroup of H.

A projectivity is called *singular* if it is not index-preserving. If a projectivity φ of a finite group G induces an index-preserving projectivity of each Sylow subgroup of G, then φ is index-preserving. Hence if φ is singular, φ induces a singular projectivity of at least one Sylow subgroup of G. To indicate the singularity of the induced projectivity on p-Sylow subgroups, we say that φ is singular at p. If a projectivity φ is singular at p, then p-Sylow subgroups of G are either cyclic or elementary abelian. We are going to distinguish two types of singularity at p. If φ is singular at p, then we say that φ has a *singularity of the first kind* at p, if there is no P-group T such that $G \supseteq T \supset S$ and S is a proper normal subgroup of T; we say that φ has a *singularity of the second kind* at p, if G has a *P-subgroup* containing S as a proper normal subgroup.

Proposition 2.7. *If φ is a singular projectivity of a finite group G, then there is a prime number p such that φ has a singularity of the first kind at p.*

Proof. By way of contradiction suppose that φ has singularities of the second kind only. By assumption φ induces a singular projectivity of a SYLOW subgroup S of G. Since φ has no singularity of the first kind, there is a P-group $T \subseteq G$ such that S is a proper normal subgroup of T. Since T is a P-group, the order of T is of the form $q^n r$ where q and r are prime numbers and $q > r$. Now $\varphi(T)$ is also a P-group. Since the greatest prime divisor of the order of a P-group is determined by its subgroup lattice, $\varphi(T)$ is of order q^{n+1} or $q^n s$ $(s < q)$. But since φ is singular at q, $\varphi(T)$ is not a q-group, but of order $q^n s$. If Q is a subgroup of T such that $\varphi(Q)$ is a q-SYLOW subgroup of $\varphi(T)$, Q is not a q-group. Hence Q contains a subgroup R of order r. R is contained in an r-SYLOW subgroup U of G. Since $\varphi(R)$ is of order $q > r$, φ is singular at r. Hence, by assumption, φ has the singularity of the second kind at r. That is, U is contained in a P-group V as a proper normal subgroup By property of P-groups r would be the greatest prime divisor of V and $\varphi(V)$. This is not the case, since q divides the order of $\varphi(V)$. This contradiction shows the validity of (2.7).

Proposition 2.8. *If φ has a singularity of the first kind at p, then G contains a normal p-SYLOW complement.*

Proof. Let S be a p-SYLOW subgroup of G. We shall prove this proposition 2.8 by using induction on the order of G. If the normalizer $N(S)$ of S is a proper subgroup, we may apply inductive hypothesis on $N(S)$. $N(S)$ is then a direct product of S and the p-SYLOW complement, and hence S is in the center of $N(S)$. By a theorem of BURNSIDE, G contains a normal p-SYLOW complement.

Assume $N(S) = G$. If G/S is not a cyclic group of prime power order, G contains at least two maximal subgroups containing S. Hence S is in the center of G by inductive hypothesis. Using a splitting theorem of SCHUR, we see that G contains a normal p-SYLOW complement. We may therefore assume that G/S is a cyclic group of order q^n (q is a prime number). Let Q be a q-SYLOW subgroup of G. By inductive hypothesis the maximal subgroup D of Q commutes elementwise with S. Hence $M = D \cup S = D \times S$ and D is a normal subgroup of G. By theorem I. 4 of p. 5, $\varphi(M) = \varphi(S) \times \varphi(D)$ and hence $\varphi(D)$ is a normal subgroup of $\varphi(G)$. Thus φ induces a projectivity of G/D onto $\varphi(G)/\varphi(D)$, which is singular at p. We consider the factor group $\overline{G} = G/D$. $\overline{M} = M/D$ is a normal and maximal subgroup of G. $\overline{M} \cong S$ and $\varphi(\overline{M})$ is not a p-group. If S is cyclic, $\Phi(\overline{G}) = N$ is a maximal subgroup of $\overline{M}$. If $\overline{G}/N$ is cyclic, then $\overline{G}$ is nilpotent and so is G. If not, $\overline{G}/N$ is a P-group, which implies that $N = e$. Hence $\overline{G}$ is a P-group. If S is not cyclic, $\overline{M}$ is completely decomposable as a G/M-module, say $\overline{M} = S_1 \times \cdots \times S_t$ where S_i is a simple G/M-module. Since φ is singular, one of the $\varphi(S_i)$, say

$\varphi(S_1)$, is not a p-group. Since S_1 is simple, S_1 does not contain any proper normal subgroup of $S_1 \cup \bar{Q}$, where $\bar{Q} = Q/D$. Hence S_1 must be one-dimensional, and $\varphi(S_1)$ is of prime order, say r. Since all subgroups of order r in $\varphi(\bar{M})$ are conjugate and generate $\varphi(\bar{M})$, we see that all subgroups of order p in $\bar{M}$ are normal. Hence G is either nilpotent or a P-group. If $\bar{G}$ were a P-group, we have $D \neq e$ by the definition of S. Since φ is singular, for some q-SYLOW subgroup Q of G, $\varphi(Q)/\varphi(D)$ would be of order p. Hence $\varphi(Q)$ would be a p-group, which contradicts the fact that $L(M) = L(S) \times L(D)$. Hence G is nilpotent and this proves the normality of Q and completes the proof.

Proposition 2.9. *If a projectivity φ of a finite group G has a singularity of the second kind at p, then*

$$G = P \times H$$

*where the orders of P and H are relatively prime, and where P is a P-group containing a p-*SYLOW *subgroup S of G as a proper normal subgroup.*

To simplify the following description we shall call a finite group G an *S-group*, if G is a direct product of a non-cyclic P-group P and a subgroup H with order relatively prime to the order of P.

Proof. Since P is a P-group and not a p-group, the order of P has the form $p^n q$ where q is a prime number $< p$. Similarly $\varphi(P)$ is of order $p^n r$ where r is a prime and $< p$. As $\varphi(S)$ is not a p-group, we may take a subgroup Q of P with order q such that $\varphi(Q)$ is of order p. Hence φ is singular at q. The proof of (2.7) indicates that φ has the singularity of the first kind at q. Hence it follows from (2.8) that G has a normal q-SYLOW complement N. It will be easy to show that Q is a q-SYLOW subgroup of G and $PN = G$.

Let K be the normalizer of S in G. Then any subgroup of K whose lattice of subgroups is isomorphic with $L(P)$ contains S. This implies that $\varphi(P)$ is a normal subgroup of $\varphi(K)$, and that P is the only P-group containing S as a proper normal subgroup. The normalizer of S in N is $N \cap K$ and there is no P-group containing S as a proper normal subgroup in N. Hence by (2.8) N contains a normal p-SYLOW complement H. H satisfies the following properties: $G = HP$, $H \cap P = e$ and the order of H is relatively prime to pq. Using induction, we shall prove that $G = H \times P$.

We take a SYLOW subgroup T of H. Then the normalizer of T in G contains some conjugate subgroup of P. Hence we may assume that H is an r-group (r is a prime). Suppose φ is singular at r. If there is no P-group containing H as a proper normal subgroup, P is normal, as it is an r-SYLOW complement of G. If H is contained in a P-group Q as a proper normal subgroup, the index $(Q:H)$ is q, and $G = Q \times S$ since H is normal in G. But this is not the case. Hence we may assume that $\varphi(H)$ is an r-group. If we take a maximal subgroup M of G containing P, $M \cap H$ is a normal subgroup of G. By inductive hypothesis

$$\varphi(M) = \varphi(P) \times \varphi(M \cap H),$$

so that $\varphi(M \cap H)$ is also a normal subgroup of $\varphi(G)$. Hence we may assume that P itself is a maximal subgroup. Since P is not cyclic, some element $\neq 1$ of P commutes with some element $\neq 1$ of H by a lemma of BURNSIDE-ZASSENHAUS (Cf. BURNSIDE [*1*], § 248). Since the subgroups of prime order in P are transformed among each other by auto-projectivities of G, any proper subgroup of P differs from its centralizer in G. Hence one of the subgroups of order p in P is normal. Hence we have $G = P \times H$ as was to be shown.

Proposition 2.10. *If φ is a singular projectivity of a finite group G, then there is a prime number p such that φ is singular at p and G contains a normal p-SYLOW complement N. If G is not an S-group, $\varphi(N)$ is a characteristic subgroup of $\varphi(G)$. (If G is an S-group, in particular if G is a P-group, $\varphi(N)$ may not be a normal subgroup.)*

Proof. The first part will follow from (2.7) and (2.8). To prove the second part, suppose that there exists an auto-projectivity ψ of G which maps N upon $M \neq N$. Since N is a p-SYLOW complement of G, some p-SYLOW subgroup, say T, of M is not e. Take a SYLOW subgroup S of G containing T. Since $\psi^{-1}(M) = N$, $\psi^{-1}(S)$ is not a p-group. On the other hand, $\psi^{-1}(S) \cap N = \psi^{-1}(T)$ and so $\psi^{-1}(T)$ is a normal subgroup of $\psi^{-1}(S)$. Hence by theorem I. 12, S is cyclic and $S = T$. Let the order of $\psi^{-1}(S)$ be q^n, and Q be a q-SYLOW subgroup of G containing $\psi^{-1}(S)$. Since G is not an S-group, there is no P-group containing Q as a proper normal subgroup and we see that G contains a normal q-SYLOW complement H_q, applying (2.9) to ψ. $\psi(Q)$ contains S. If S is a proper subgroup of $\psi(Q)$, S is a q-SYLOW complement of $\psi(Q)$ and so normal. This is impossible. Hence $\psi(Q) = S$. Suppose S is mapped upon p-, q-, . . ., r-SYLOW subgroups of G by all auto-projectivities of G. Then G contains normal p-, q-, . . ., r-SYLOW complements $H_p = N$, $H_q, \ldots, H_r$. The intersection $D = H_p \cap H_q \cap \cdots \cap H_r$ is invariant under all auto-projectivities of G, and the factor group G/D is cyclic. Hence $\varphi(H_p)$, $\varphi(H_q), \ldots, \varphi(H_r)$ are characteristic subgroups of $\varphi(G)$. This completes the proof.

We have now obtained the main results of our discussion (SUZUKI [*1*]).

Theorem 8. *Let G be a finite group. If G contains no normal* SYLOW *complement S whose factor group G/S is either cyclic or a P-group, then every projectivity of G is index-preserving.*

Theorem 8 is an easy consequence of (2.10). As special cases of this theorem we know that *every projectivity of a perfect group, or of the symmetric group of n letters* ($n \geq 4$), *is index-preserving.* For perfect groups, we have furthermore

Theorem 9. *Every projectivity maps finite perfect groups onto finite perfect groups.*

Proof. If H is not perfect, H contains a maximal subgroup N which is normal. By theorem 8, φ is index-preserving, so is φ^{-1}. Hence by (2.6) of p. 42 $\varphi^{-1}(N)$ is a normal subgroup of G. Since N is maximal, $G/\varphi^{-1}(N)$ is cyclic, which contradicts the assumption of perfectness. Hence H is perfect.

Using induction we may prove easily the following theorem SUZUKI [*1*], ZAPPA [*9*]).

Theorem 10. *Every projectivity maps finite solvable groups upon finite solvable groups.*

Theorem 11. *Let φ be a projectivity of a finite group G. G contains a normal subgroup N such that φ induces an index-preserving projectivity of N, $\varphi(N)$ is normal in $\varphi(G)$ and the factor group G/N is a direct product of a cyclic group and several P-groups of relatively prime order.*

Proof. We decompose G into a direct product of several P-groups and a subgroup H: $G = P_1 \times \cdots \times P_r \times H$, where any two factors have relatively prime orders, each P_i is a P-group, and H is not an S-group. If φ induces a singular projectivity of H, H contains a normal SYLOW complement N_1. By (2.10), $\varphi(N_1)$ is a characteristic subgroup of $\varphi(H)$ and one of the "exceptional" SYLOW subgroups is excluded from N_1. Repeating this process, we see that H contains a normal subgroup N such that φ induces an index-preserving projectivity of N and $\varphi(N)$ is also normal. Since φ is singular at every prime divisor of the order of H/N, H/N is a direct product of a cyclic group and several P-groups. Since N is a normal subgroup of G, N satisfies all properties of this theorem. If φ induces an index-preserving projectivity of H, we may take $N = H$.

Corollary 1. *Every projectivity of a finite group induces an index-preserving projectivity of the second commutator subgroup.*

Corollary 2. *Let φ be a projectivity of a finite group G. If G is not an S-group, G contains a normal subgroup N such that φ induces an index-preserving projectivity of N, $\varphi(N)$ is normal in $\varphi(G)$, and G/N is cyclic.*

Proof. By theorem 11, there is a normal subgroup N_0 such that $\varphi(N_0)$ is a normal subgroup of $\varphi(G)$ with the same order as N_0, and

G/N_0 is a direct product of a cyclic group and several P-groups. Since G is not an S-group, we see from the above proof that G/N_0 is abelian. Let $\varphi(G)/\varphi(N_0)$ be a direct product $P_1 \times \cdots \times P_t \times Z$ where the P_i are P-groups and Z is cyclic. Since φ is singular at every prime divisor of the order of G/N, each P_i is non-abelian. Each P_i contains a maximal normal subgroup S_i. Let N be the subgroup of G such that $\varphi(N)/\varphi(N_0) = S_1 \times \cdots \times S_t$. Then it is easy to show that N meets the requirements of this corollary.

Corollary 3. *The orders of two finite groups with isomorphic lattices of subgroups contain the same number of prime divisors.*

Theorem 12. *If a finite group G admits an auto-projectivity ψ such that ψ maps a* Sylow *subgroup onto a subgroup which is not a* Sylow *subgroup, then G is an S-group.*

Proof. ψ or ψ^{-1} map a Sylow subgroup of G upon a subgroup which is not of prime power order. We may assume that $\psi(S)$ is not of prime power order for some p-Sylow subgroup S. If G is not an S-group, G contains a p-Sylow complement N. By assumption $\psi(S) \cap N \neq e$, or $S \cap \psi^{-1}(N) \neq e$. Hence ψ^{-1} transforms N onto another group M. We see from the proof of (2.10) in p. 45 that $\psi(S)$ is a Sylow subgroup of G. This contradiction shows that G is an S-group.

7. The images of normal subgroups under projectivities of finite groups.

Let G be a finite group and U a subgroup of G. The intersection of all normal subgroups of U whose factor groups are p-groups is denoted by $N_p(U)$. The intersection $N(U) = \cap_p N_p(U)$ for all p is the smallest normal subgroup with a nilpotent factor group. We write $N^1(U) = N(U)$ and $N^{i+1}(U) = N(N^i(U))$. Then the intersection $S(U) = \cap_i N^i(U)$ is the smallest normal subgroup with a solvable factor group.

Proposition 2.11. *Let M be a maximal subgroup of a group G. Assume that M is normal and $(G:M) = p$. If a projectivity φ of G is not singular at p, $\varphi(M)$ is a normal subgroup of $\varphi(G)$.*

This may be proved in a similar fashion as (2.6) of p. 42.

Proposition 2.12. *If φ is a projectivity of a finite group G, and if $G/N_p(G)$ is not a P-group, then $\varphi(N_p(G)) = N_q(\varphi(G))$ for some q.*

Proof. We prove this proposition by induction on the order. Let $H = N_p(G)$. If φ is singular at p, a p-Sylow subgroup S is cyclic, since $G/N_p(G) \cong S$. Hence from (2.10), p. 45, we conclude that $\varphi(H) = N_q(\varphi(G))$. We may therefore assume that φ is non-singular at p. Take a maximal subgroup M containing H. Since M is a normal subgroup, $\varphi(M)$ is also normal by (2.11). H coincides with $N_p(M)$. Hence if M/H is not a P-group, $\varphi(H) = N_q(\varphi(M))$ by the inductive hypothesis. But φ is not singular at p, $q = p$ and $\varphi(H) = N_p(\varphi(G))$. If M/H is a P-group, we may take maximal subgroups U_i of M containing H such that $\bigcap U_i = H$. Since φ is non-singular at p, each $\varphi(U_i)$ is normal in $\varphi(M)$, and so is $\varphi(H) = \bigcap_i \varphi(U_i)$. Hence $N_p(\varphi(M)) \subseteq \varphi(H)$. On the other hand, if we take a subgroup V of G such that $\varphi(V) = N_p(\varphi(M))$, V is a normal subgroup of M and M/V is a p-group. Hence $V = N_p(M) = H$ or $\varphi(H) = \varphi(V) = N_p(\varphi(M))$. Hence we have $N_p(\varphi(M)) = \varphi(H)$, which implies that $\varphi(H)$ is a normal subgroup of $\varphi(G)$ and $\varphi(H) = N_p(\varphi(G))$.

If the factor group $G/N_p(G)$ is a P-group, the situation is rather different. $\varphi(N_p(G))$ is not always self-conjugate in $\varphi(G)$. Put $H = N_p(G)$, and suppose that $\varphi(H)$ is not a characteristic subgroup of $\varphi(G)$. Then there exists an automorphism σ of $\varphi(G)$ which maps $\varphi(H)$ onto $U, \neq \varphi(H)$. This σ induces an auto-projectivity of $L(\varphi(G))$, and hence of $L(G)$. We denote this induced auto-projectivity of G by the same letter σ. For this σ we prove

Proposition 2.13. *σ^{-1} is singular at some prime q and each q-*Sylow *subgroup is contained in a P-group as a proper normal subgroup.*

Proof. If a projectivity φ is not singular at p, $\varphi(H)$ is a normal subgroup of $\varphi(G)$ by (2.11). Hence if the induced auto-projectivity σ of G were not singular at p, H would be invariant under σ. Hence σ is singular at p. Since $G \neq N_p(G)$, H must be a p-Sylow complement by (2.8) and (2.11). Since $H \neq H^\sigma$, and since σ is induced by an automorphism of $\varphi(G)$, we conclude from (2.10) that G is an S-group and σ^{-1} satisfies the conditions of this proposition.

Proposition 2.14. *Let U be a subgroup of G, V be a normal subgroup of U and φ be a projectivity of G. Assume that $\varphi(V)$ is a normal subgroup of $\varphi(U)$. If $\varphi(N_p(V))$ is not a normal subgroup of $\varphi(U)$, then G is an S-group.*

Proof. By (2.12) $V/N_p(V)$ is a P-group. By hypothesis there exists an inner automorphism of $\varphi(G)$ which maps $\varphi(U)$ upon itself but $\varphi(N_p(V))$ upon a subgroup different from $\varphi(N_p(V))$. The induced auto-projectivity ψ of G is singular at p. We may apply (2.13) onto ψ^{-1} and conclude that G is an S-group.

Theorem 13. *Let $S(G)$ be the smallest normal subgroup with solvable factor group of a finite group G. Then $\varphi(S(G)) = S(\varphi(G))$ for every projectivity φ of G.*

Proof. Suppose G is not an S-group. By definition of $S(G)$, there is a series of subgroups $H_0 = G \supset H_1 \supset \cdots \supset H_r = S(G)$ such that

$H_i = N_{p_i}(H_{i-1})$ for some p_i $(i = 1, 2, \ldots, r)$. By (2.12) and (2.14) each $\varphi(H_i)$ is a normal subgroup of $\varphi(G)$. Each factor group $\varphi(H_{i-1})/(H_i)$ is either a p-group or a P-group, and so solvable. Hence $\varphi(G)/\varphi(S(G))$ is solvable. Hence we have $\varphi(S(G)) \supseteq S(\varphi(G))$. Considering φ^{-1}, we get $S(G) \subseteq \varphi^{-1} S(\varphi(G))$ or $\varphi(S(G)) \subseteq S(\varphi(G))$. Hence $\varphi(S(G)) = S(\varphi(G))$. In general G is a direct product of several P-groups and a subgroup H which is not an S-group. Since $S(G) = S(H)$, and $S(\varphi(G)) = S(\varphi(H))$, we get $\varphi(S(G)) = \varphi(S(H)) = S(\varphi(H)) = S(\varphi(G))$.

Proposition 2.15. *Let φ be a projectivety of a finite group G, and H a normal subgroup of G. If $\varphi(H)$ is not normal in $\varphi(G)$, there exist three normal subgroups D, N and M of G such that*

(1) $N \supseteq H \supseteq M$, $D \supseteq M$ *and* $D \cup H \neq G$,

(2) $\varphi(D)$, $\varphi(N)$ *and* $\varphi(M)$ *are self-conjugate in* $\varphi(G)$,

(3) *the factor groups* G/D *and* N/M *are both solvable, and*

(4) *if G is not an S-group, some prime number p divides the indices* $(G:D)$, $(N:H)$ *and* $(H:M)$.

Proof. Suppose first that this proposition has already been proved for groups which are not S-groups. If G is an S-group, then G is a direct product of a group U and a group V where U is a direct product of several P-groups and V is not an S-group. $H \cap V$ is a normal subgroup of V. If $\varphi(H \cap V)$ is normal in $\varphi(V)$, we may take $D = V$, $N = U(H \cap V)$, and $M = H \cap V$. If $\varphi(H \cap V)$ is not normal in $\varphi(V)$, there are normal subgroups D_0, N_0 and M_0 with properties (1) to (3). Then $D = D_0$, $N = N_0 \cup U$ and $M = M_0$ have the required properties.

Assume now that G is not an S-group. We prove this proposition by induction on the order of the factor group G/H. If an auto-projectivity ψ of G is singular at some prime p, the p-Sylow subgroups of G are cyclic, G contains a self-conjugate p-Sylow complement, and ψ maps a p-Sylow subgroup upon a Sylow subgroup. If H is a maximal subgroup of G, H contains $N_p(G)$ for $p = (G:H)$. Hence the assertion follows from (2.12) and (2.13), if we set $N = G$ and $D = M = N_p(G)$. Take a subgroup T of G such that $T \supseteq H$ and T/H is a Sylow subgroup of G/H. Then $S(H) = S(T)$. Hence $\varphi(S(H)) = \varphi(S(T)) = S(\varphi(T))$ by theorem 13. This implies that $\varphi(S(H))$ is a normal subgroup of $\varphi(T)$. Since this is true for every Sylow subgroup T/N, $\varphi(S(H))$ is a normal subgroup of $\varphi(G)$. We take a largest subgroup M of G such that $H \supseteq M \supseteq S(H)$, and both M and $\varphi(M)$ are self-conjugate. Since $H \supseteq M \supseteq S(H)$, H/M is a solvable group. Hence H/M contains a normal subgroup P/M of G/M such that P/M is a p-group. We may assume that P is maximal under these restrictions. Then P/M is the intersection of all p-Sylow subgroups of H/M. Take any subgroup T

of G such that $T \supseteq H$ and T/H is a q-SYLOW subgroup of G/H ($p \neq q$). Then P/M is also the intersection of all p-SYLOW subgroups of T/M. Hence if $\varphi(P)$ were not self-conjugate in $\varphi(T)$, G would admit an auto-projectivity ψ which would be singular at p. Hence G would contain a self-conjugate p-SYLOW complement U. $M(T \cap U)/M$ would be a p-SYLOW complement of T/M and $\varphi(M(T \cap U)) = \varphi(M) \cup (\varphi(T) \cap \varphi(U))$ would be a normal subgroup of $\varphi(T)$. Since ψ is induced by an inner automorphism of $\varphi(T)$, ψ leaves invariant $M(T \cap U)$. This is a contradiction because ψ would not map p-SYLOW subgroups onto p-SYLOW subgroups. Hence $\varphi(P)$ is normal in $\varphi(T)$. If we consider all subgroups T such that $T \supseteq H$ and T/H are q-SYLOW subgroups, and if we let q run through all prime divisors of $(G:H)$ except p, $\varphi(P)$ is normal in the union of all such $\varphi(T)$. By the definition of M, $\varphi(P)$ is not normal in $\varphi(G)$. Hence the union of all such T's is not G, which implies that $H \cup N_p(G) \neq G$. Set $D = M \cup N_p(G)$, then $H \cup D = H \cup N_p(G) \neq G$.

Since P is a normal subgroup of G, every p-SYLOW subgroup of G/M contains P/M. Let Q/M be the intersection of all p-SYLOW subgroups of G/M. Q is a normal subgroup of G. If $\varphi(Q)$ were not self-conjugate G would admit a singular auto-projectivity. In the same way as before we get a contradiction. Hence $\varphi(Q)$ is a normal subgroup of $\varphi(G)$. This implies, in particular, that $Q \neq P$. Since $P = Q \cap H$, $Q\,H$ is not equal to G, and $Q\,H/H \cong Q/Q \cap H = Q/P$ is a p-group. We may apply the inductive hypothesis on $Q\,H$ and see that there is a normal subgroup N of G such that $N \supseteq Q\,H$, $N/Q\,H$ is solvable and $\varphi(N)$ is normal. Since $Q\,H \supseteq H$ and $Q\,H/H$ is solvable, N/M is a solvable group. From the above proof we can easily see that p divides the indices $(G:D)$, $(N:H)$ and $(H:M)$. This completes the proof.

Theorem 14. *Let N be a normal subgroup of a finite group G, and φ be a projectivity of G. $\varphi(N)$ is a normal subgroup of $\varphi(G)$ if one of the following conditions is satisfied:*

(1) *G/N is perfect,*

(2) *G/N contains no proper solvable normal subgroup,*

(3) *N is perfect, or*

(4) *G is not an S-group, and the order of N is relatively prime to the index $(G:N)$.*

Corollary 1. *Let $R(G)$ be the radical of G,* i. e. *the maximal solvable normal subgroup of G. Then $\varphi(R(G)) = R(\varphi(G))$ for any projectivity of G.*

Proof. $\varphi(R(G))$ is normal by theorem 14, and solvable by theorem 10, p. 46. Hence $\varphi(R(G)) \subseteq R(\varphi(G))$. Considering φ^{-1}, we get this corollary.

Corollary 2. Let $G = G_0 \supset G_1 \supset \cdots \supset G_r = e$ be a composition series of a finite group G. If G is not an S-group,

$$G = \psi(G_0) \supset \psi(G_1) \supset \cdots \supset \psi(G_r) = e$$

is also a composition series of G for any auto-projectivity ψ of G.

Proof. If $\psi(G_i)$ is not a normal subgroup of $\psi(G_{i-1})$, the index $(G_{i-1}:G_i)$ is a prime number p, and ψ is singular at p. Hence G is an S-group by (2.14).

Theorem 15. *Let G be a perfect group of finite order. Then for any projectivity φ of G, $\varphi(G)$ is a perfect group of the same order as G and φ induces an isomorphism of the lattice of all normal subgroups of G onto that of $\varphi(G)$. If we denote the center of G by Z, $\varphi(Z)$ is the center of $\varphi(G)$.*

Proof. The first assertion follows from theorems 8 and 9 of p. 45. The second one is a consequence of theorem 14. The last one is proved as follows. Take a p-Sylow subgroup S of Z, and a q-Sylow subgroup Q of G. If $p \neq q$, then $S \cup Q = S \times Q$. Hence $\varphi(S \cup Q) = \varphi(S) \times \varphi(Q)$ by theorem I. 2. Hence $\varphi(S)$ is elementwise permutable with $\varphi(Q)$. Since G is perfect, $G = \bigcup Q$ and hence $\varphi(S)$ is in the center of $\varphi(G)$. This shows that $\varphi(Z)$ is a part of the center. Considering φ^{-1}, we conclude that $\varphi(Z)$ is the center of $\varphi(G)$.

8. The number of finite groups with given lattice of subgroups.

Let L be a finite lattice. The number of groups whose subgroup lattices are isomorphic with L may vary from zero to infinity. We can, however, prove the following theorem (Suzuki [*1*], p. 350, theorem 1).

Theorem 16. *If L has no chain as its direct component, then the number of non-isomorphic groups whose lattices of subgroups are isomorphic to L is finite.*

Proof. By theorem I. 4, p. 5, we may assume that L is irreducible and not a chain. We denote by n the dimension of L and by m the number of elements in L. Let G be any group whose subgroup lattice is isomorphic to L. Then G is a finite group. We shall prove that any prime divisor p of the order of G is smaller than m. If this has been done it follows that the order of G must be smaller than m^{nm}. Hence the types of G are finite in number.

Consider now a p-Sylow subgroup S of G. If $G = S$, G is not cyclic since $L = L(G)$ is not a chain. Hence p is smaller than m. Assume that S is not a normal subgroup of G. Then the number of conjugate subgroups of S is conjugate to 1 modulo p. Hence $p < m$. If $S \neq G$ is a normal subgroup of G, G has a p-Sylow complement H by the splitting theorem of Schur. H is not self-conjugate, since otherwise $L(G)$ would be directly decomposable by theorem I. 4, of p. 5. The number of conjugate subgroups of H is now a power of p. Hence we get again $p < m$.

9. The group of auto-projectivities.

Let G be a group. The totality of automorphisms of its subgroup lattice forms a group, which we denote by $A(G)$. The index-preserving auto-projectivities of G form a subgroup $J(G)$ of $A(G)$. $J(G)$ is not always a normal subgroup of $A(G)$. An example may be supplied by the symmetric group of three letters. But we can prove the following result.

Theorem 17. *Let G be a finite group. If G is not an S-group, $J(G)$ is a normal subgroup of $A(G)$.*

Proof. By assumption and (2.10) of p. 45 we conclude that any auto-projectivity of G maps Sylow subgroups onto Sylow subgroups. Assume a p-Sylow subgroup S is mapped upon p-, q-, . . ., r-Sylow subgroups of G. Then by (2.8) and (2.9) on pp. 43—44 G contains p-, q-, . . ., r-Sylow complements $H_p, \ldots, H_r$ and the intersection H of all these $H_p, \ldots, H_r$ is invariant under all auto-projectivities of G. Hence every auto-projectivity ψ of G induces an auto-projectivity $\overline{\psi}$ of G/H and the mapping $\psi \rightarrow \overline{\psi}$ defines a homomorphism of $A(G)$, and $J(G)$ is the kernel of this homomorphism. Hence $J(G)$ is a normal subgroup of G.

The structure of the factor group $A(G)/J(G)$ is not yet known. One of the important subgroups of $J(G)$ is the subgroup $I(G)$ consisting of all projectivities induced by group automorphisms. $I(G)$ is not always self-conjugate in $J(G)$. An example is supplied by the non-abelian group of order 10. We shall consider the structure of $I(G)$.

Let K be the subgroup of G consisting of all elements which commute with every subgroup of G. K is called the *norm* of G. Norms are characteristic subgroups and are abelian or Hamiltonian. Clearly K contains the center Z of G.

Proposition 2.16. *If G is a finite group such that $Z = e$, then $K = e$.* (Baer [*2*], p. 254.)

Proof. By definition K is a normal subgroup of G. Consider a p-Sylow subgroup K_p of K. If K_p is not equal to e, K_p contains a central element a, $\neq 1$, of a p-Sylow subgroup S_p of G. Since $a \in K_p$, a commutes with every q-Sylow subgroup S_q of G. Hence if $p \neq q$, a commutes with every element of S_q. Hence a is in the center of G.

We proved actually that, if p divides the order of the norm of G, then p divides the order of the center. Hence if we define the hypernorm in a similar way as the hypercenter, the hypernorm of a finite group coincides with the hypercenter. It is not yet known whether the same propositions are true or false for infinite groups.

Theorem 18. *If the center of a finite group G consists of the identity element only, then I (G) is isomorphic with the group of all automorphisms of G, and hence any auto-projectivity of G is induced by at most one group automorphism.*

Proof. Let A be the group of all automorphisms of G. If we denote by φ_σ the auto-projectivity induced by σ, the mapping $\sigma \to \varphi_\sigma$ defines a homomorphism of A onto $I(G)$. Let N be the kernel of this homomorphism. A contains the group G of inner automorphisms and $K = G \cap N$ is the norm of G. Hence by assumption and (2.16) $G \cap N = e$, so that every element of N commutes with all elements of G. If we take $\tau \in N$, then for any pair of elements g and t of G, we have $t\, g^\tau\, t^{-1} = t^\tau\, g^\tau\, t^{-\tau}$. Hence $t^{-1}\, t^\tau$ is in the center of G. Hence $t^\tau = t$, i.e. $N = e$. This means that $\sigma \to \varphi_\sigma$ is an isomorphism of A onto $I(G)$.

Let ψ be an auto-projectivity of a finite group G, N be a normal subgroup of G. $\psi(N)$ is not necessarily a normal subgroup. Assume now that ψ satisfies the following conditions:

ψ is not singular at p and for every normal subgroup U of any p-Sylow subgroup S of G, $\psi(U)$ is normal in $\psi(S)$.

An auto-projectivity satisfying the above conditions is called *locally normal at p*. If an auto-projectivity ψ is locally normal at every prime divisor p of the index of the commutator subgroup, then we may show using (2.8), (2.9) and (2.15) that $\psi(N)$ is a normal subgroup whenever N is a normal subgroup of G. This proposition reduces the consideration to the local one, i. e. essentially to the case of p-groups. But the situation is not yet clear even if G itself is a p-group.

10. Projectivities of simple groups.

Let G be a non-abelian simple group of finite order. If there exists a projectivity of G upon another group H, H is also a non-abelian simple group with the same order as G. It is not yet known whether H is necessarily isomorphic with G or not. We can prove however the following result (Suzuki [*1*], p. 365, theorem 23).

Theorem 19. *If G is a finite non-abelian simple group, and if* $L(G \times G) \cong L(H)$, *then* $H \cong G \times G$.

Proof. We shall take isomorphic copies A and B of G, and denote $G \times G$ by $K = A \times B$. Let φ be the projectivity of K onto H. By theorem 14, p. 50, both $\varphi(A)$ and $\varphi(B)$ are normal in H, so that we have $H = \varphi(A) \times \varphi(B)$.

Suppose in general a group L is a direct product of two subgroups M and N. A subgroup U of L is called a *C-subgroup* of L, if the conditions

$$U M = U N = L \text{ and } U \cap M = U \cap N = e$$

are satisfied. If U is a C-subgroup of L, then every element u of U is a pair (m, n) of $m \in M$ and $n \in N$, and the projections $u \to m$ and $u \to n$ of U into M and N are homomorphisms. Moreover, since U is a C-subgroup, these projections are isomorphisms onto M or N. Hence a C-subgroup consists of all elements of L of the form $(m, \sigma(m))$ where σ is an isomorphism of M onto N, and vice versa.

Returning to our case, a subgroup U of K is a C-subgroup if and only if $\varphi(U)$ is a C-subgroup of H. This implies in particular that $\varphi(A) \cong \varphi(B)$.

Let U and V be two C-subgroups of K. Suppose U consists of the elements $(a, \sigma(a))$ and V of the elements $(a, \tau(a))$. Then $\sigma\tau^{-1}$ is an automorphism of B, which may be denoted by $\sigma(U; V)$. If U, V, W are C-subgroups of K, then

$$\sigma(U; V)\,\sigma(V; W) = \sigma(U; W).$$

To every C-subgroup U of K and every automorphism σ of B there exists one C-subgroup V such that $\sigma(U; V) = \sigma$. Let U, V, W and T be C-subgroups of K such that

$$\sigma(U; V) = \sigma(W; T) = \sigma.$$

We shall prove that T is determined by U, V, W and $L(K)$. The mapping $(a, \sigma(b)) \to (a, b)$ defines an automorphism $\bar{\sigma}$ of K. $\bar{\sigma}$ induces an auto-projectivity φ_σ of K, which maps U onto V, W onto T and which leaves invariant B and all subgroups of A. If the auto-projectivity ψ of K satisfies

$$\psi(U) = V,\ \psi(B) = B \text{ and } \psi(X) = X$$

for every subgroup X of A, then ψ maps W upon T. In order to prove this assertion, let $\pi = \psi^{-1}\varphi_\sigma$. Then π leaves invariant U, B and all subgroups of A. Take any subgroup Y of U. Then

$$Y = Y B \cap U \text{ and } Y B = B \cup (Y B \cap A).$$

Hence applying π we know that $\pi(Y) = Y$. Similarly for any subgroup Z of B, we have $\pi(Z) = Z$, since

$$Z = Z A \cap B \text{ and } Z A = A \cup (Z A \cap U).$$

π induces thus the identical auto-projectivity of B. Let Z be any C-subgroup of K. Then $\pi(Z)$ is a C-subgroup and $\tau = \sigma(\pi(Z); Z)$ is an automorphism of B which induces an auto-projectivity φ_τ of K. φ_τ maps $\pi(Z)$ upon Z, and hence $\varphi_\tau\,\pi$ induces the identical auto-projectivity of B. On the other hand $\varphi_\tau \pi$ is induced by an automorphism τ of B on B. By theorem 18, τ must be the identity. Hence we get $\pi(Z) = Z$, which implies

$$\psi(W) = \varphi_\sigma(W) = T.$$

Since H is also a direct product of two simple groups, we may define $\sigma(U'; V')$ in H. The above considerations show that we have

$\sigma(U; V) = \sigma(W; T)$ if and only if $\sigma\big(\varphi(U); \varphi(V)\big) = \sigma\big(\varphi(W); \varphi(T)\big)$.

Hence the mapping

$$f: \sigma(U; V) \to \sigma\big(\varphi(U); \varphi(V)\big)$$

is well defined and since $\sigma(U; V)\,\sigma(V; W) = \sigma(U; W)$, f is an isomorphism of the group of all automorphisms of B onto that of $\varphi(B)$. Since B is simple, we conclude that B is isomorphic with $\varphi(B)$.

The next theorem is a simple consequence of theorems 13 and 19.

Theorem 20. *Let G be a simple group. G is isomorphic with H if and only if the lattices of subgroups $L(G \times G)$ and $L(H \times H)$ are isomorphic.*

Remark. In theorem 19, the given projectivity φ of $G \times G$ onto H may *not* be induced by any group isomorphism.

11. Characteristic chains of subgroup lattices.

An element of a lattice is called *characteristic* if it is left invariant by all automorphisms of the lattice. The smallest and the greatest elements are characteristic. A characteristic element is *proper*, if it is neither I nor O. If a lattice L contains no proper characteristic element, then L is called *characteristic simple.*

Proposition 2.17. *If the subgroup lattice $L(G)$ of a finite group G is characteristic simple, then G belongs to one of the following types:* (1) *a P-group,* (2) *a cyclic group of square-free order,* (3) *a characteristic simple group, or* (4) *a direct product of non-isomorphic non-abelian simple groups with isomorphic subgroup lattices.*

Proof. Consider a maximal normal subgroup N of G. If $N = e$, G is of type (3). Hence we may assume $N \neq e$, and distinguish two cases.

Case I. $(G:N) = p$ is a prime number. Since $N \neq e$, some autoprojectivity ψ maps N onto $\psi(N) \neq N$. If $\psi(N)$ is not normal, ψ is singular at p by (2.11). Hence by (2.12) of p. 47 G is a P-group. If $\psi(N)$ is a normal subgroup of index p, G is again a P-group by (2.12). If $\psi(N)$ is normal, but $(G:\psi(N)) = q \neq p$, then $G/N_p(G)$ is cyclic. Since $\psi(N_p(G)) = N_q(G)$, G is a cyclic group of square-free order.

Case II. G/N is non-abelian. By theorem 14, p. 50, $\psi(N)$ is normal for every auto-projectivity ψ of G. Hence G is a direct product of non-abelian simple groups G_i ($i = 1, 2, \ldots, r$). If two components, say G_1 and G_2, are isomorphic, then by theorem 19,

$$G_1 \times G_2 \cong \psi(G_1 \times G_2).$$

Hence all components must be isomorphic to each other.

Remark. It is not known whether the groups of type (4) actually exist.

We may define a characteristic chain in subgroup lattices. A series of subgroups

$$e = U_0 \subset U_1 \subset \cdots \subset U_r = G$$

is called a *characteristic chain* if each U_i is a characteristic element of $L(G)$, and each interval U_i/U_{i-1} is characteristic simple for every $i = 1, 2, \ldots, r$.

Theorem 21. *A finite group G is solvable if and only if G has a characteristic chain*

$$e = U_0 \subset U_1 \subset \cdots \subset U_r = G$$

such that each interval U_i/U_{i-1} is modular.

This is a consequence of (2.17), and contains theorem 10 of p. 46 as a corollary. p-groups and P-groups may be characterized similarly.

Theorem 22. *A finite group G is a p-group or a P-group, if its subgroup lattice $L(G)$ is a lower semi-modular lattice, satisfying*

(1) *every interval is irreducible, and*

(2) *the number of atoms in any ideal of $L(G)$ is $\equiv 1 + p \pmod{p^2}$. The last condition may be replaced by*

(2)′ *the number of chains of dimension 2 in any ideal of $L(G)$ is $\equiv 0 \pmod{p}$.*

If $p > 2$, the converse is also true. The conditions (2) and (2)′ are theorems of KULAKOFF and MILLER respectively.

12. Representation of lattices as subgroup lattices.

No sufficient condition under which a lattice L is isomorphic with a subgroup lattice is known. P. WHITMAN [*1*] proved the following theorem.

Theorem 23. *Every lattice is isomorphic with a sublattice of a suitable subgroup lattice, and any relations which are satisfied by every subgroup lattice are trivial.*

13. The situation-preserving mappings.

A projectivity φ of a group G is called *situation-preserving* if the following conditions are satisfied:

(1) φ is strictly index-preserving, and

(2) subgroups U and V are conjugate in a subgroup W of G, if and only if $\varphi(U)$ and $\varphi(V)$ are conjugate in $\varphi(W)$.

Such mappings were first considered by ROTTLÄNDER [*1*]. For situation-preserving projectivities of abelian groups and Hamiltonian groups, see BAER [*1*, *2*]. HONDA [*1*] studied the situation-preserving projectivities of a finite group with cyclic SYLOW subgroups.

Few results are known for general situation-preserving projectivities. Two groups are not necessary isomorphic even if there exists a situation-preserving projectivity between them. Here we shall give an example due to ROTTLÄNDER.

Let $G(\mu)$ be the group of order $p^2 q$ (p, q are primes, $p > q$) generated by three elements a, b and c with the relations:

$$a^p = b^p = c^q = 1, \quad a\,b = b\,a$$

$$c\,a\,c^{-1} = a^r, \quad c\,b\,c^{-1} = b^{r^\mu}$$

where $r \not\equiv 1$, $r^q \equiv 1 \pmod p$ and $\mu \not\equiv 0, 1 \pmod q$. Then

$$G(\mu) \cong G(\nu) \text{ if and only if } \mu\,\nu \equiv 1 \pmod q,$$

but there exists a situation-preserving mapping of $G(\mu)$ onto $G(\nu)$ for every pair (μ, ν).

Chapter III.

Homomorphisms of subgroup lattices.

1. The kernels of a homomorphism of a subgroup lattice.

A homomorphic mapping of the subgroup lattice $L(G)$ of a group G onto a lattice L is called an *L-homomorphism* of G onto L. Thus an L-homomorphism φ is a single valued mapping from $L(G)$ onto L such that

$$\varphi(U \cap V) = \varphi(U) \cap \varphi(V)$$

and

$$\varphi(U \cup V) = \varphi(U) \cup \varphi(V).$$

By an obvious inductive argument we see that

$$\varphi(\cap_\lambda U_\lambda) = \cap_\lambda \varphi(U_\lambda)$$

and

$$\varphi(\cup_\lambda U_\lambda) = \cup_\lambda \varphi(U_\lambda)$$

hold for any finite number of subgroups U_λ ($\lambda \in \Lambda$). If these relations hold for any (finite or infinite) number of subgroups, φ is called *complete*.

The intersection of all subgroups U which are mapped by φ onto the greatest element I of L is called the *upper kernel* or shortly *u-kernel* of φ. φ does not, in general, map the u-kernel onto I. For instance, the mapping φ of the subgroup lattice of an infinite cyclic group Z defined by

$$\varphi(e) = 0, \ \varphi(U) = I \text{ if } U \neq e,$$

is an L-homomorphism of Z onto a 2-dimensional chain, and its u-kernel is equal to e. If, however, φ is complete, φ always maps the u-kernel onto I. More generally, if φ is complete, there exist to every $a \in L$ a greatest and a least subgroup of G which are mapped onto a by φ.

Theorem 1. *The u-kernel of φ is a normal subgroup of G.* (SUZUKI [2], p. 375, theorem 3, SATO [5], p. 109, theorem 2.)

Proof. It is sufficient to prove that $\varphi(U) = I$ implies $\varphi(g\,U\,g^{-1}) = I$ for every $g \in G$. Put $\{g\} = Z$. Then

$$\varphi(U \cap Z) = \varphi(U) \cap \varphi(Z) = \varphi(Z),$$

since $\varphi(U) = I$. Now $\{g\,U\,g^{-1}, g\} \supseteq U$, and hence we have

$$I = \varphi\{g\,U\,g^{-1}, g\} = \varphi(g\,U\,g^{-1}) \cup \varphi(Z).$$

On the other hand,

$$g\,U\,g^{-1} \supseteq g(U \cap Z)\,g^{-1} = U \cap Z$$

which implies that

$$\varphi(g\,U\,g^{-1}) \supseteq \varphi(U \cap Z) = \varphi(Z).$$

Hence we conclude $I = \varphi(g\,U\,g^{-1})$, q. e. d.

The *lower kernel* or the *l-kernel of* φ is defined as the dual of the u-kernel: the l-kernel is the join of all subgroups of G which are mapped onto the least element O of L. The l-kernel of φ is then the totality of elements g of G satisfying $\varphi\{g\} = O$.

Theorem 2. *The l-kernel of φ is a normal subgroup of G.* (SUZUKI [2], p. 375, theorem 2, SATO [5], p. 110, theorem 1.)

Proof. Let U be a subgroup of G such that $\varphi(U) = O$. Take an element g of G and let $Z = \{g\}$. Then

$$\varphi(U \cup Z) = \varphi(U) \cup \varphi(Z) = \varphi(Z).$$

Since $g\,U\,g^{-1} \subseteq U \cup Z$, $\varphi(g\,U\,g^{-1}) \subseteq \varphi(Z)$. On the other hand,

$$\varphi(g\,U\,g^{-1} \cap Z) = \varphi\big(g(U \cap Z)\,g^{-1}\big) = \varphi(U \cap Z) = 0.$$

Hence $\varphi(g\,U\,g^{-1}) = O$. The theorem follows easily.

2. Complete L-homomorphisms onto cyclic groups.

Complete L-homomorphisms onto cyclic groups have been investigated by WHITMAN [2], ZAPPA [6] and SUZUKI [2].

Proposition 3.1. *A group G admits a complete L-homomorphism onto a finite chain of dimension n ($n > 0$), if and only if G is a torsion group which contains normal subgroups G_0 and E such that*

(1) *G_0 is a cyclic group of prime power order p^m,*

(2) $(G_0 : G_0 \cap E) \geq p^n$,

(3) *every element of p-power order of G generates a subgroup which either contains or is contained in G_0,*

(4) *the factor group G/E is a p-group, and*

(5) *every element of G_0 commutes with every element of E.*

Proof. Let C be a finite chain of dimension n, and φ be a complete L-homomorphism of G onto C. Denote by G_0 the u-kernel, and by E the l-kernel of φ. Then G_0 and E are normal subgroups of G by theorems 1 and 2. Since φ is complete, every subgroup of G_0 is contained in a maximal subgroup, and every maximal subgroup of G_0 is mapped upon the maximal element of C. Hence G_0 contains only one maximal subgroup, so that G_0 is a cyclic group of prime power order, say of order p^m. This proves (1). φ maps G_0 onto C, and the l-kernel of this induced L-homomorphism of G_0 is exactly $E \cap G_0$. Hence the index $(G_0 : G_0 \cap E)$ is not smaller than p^n, and (2) is proved. Let $G_0 = \{a\}$. Since G_0 is a finite normal subgroup of G, the centralizer Z of G_0 is of finite index in G. Suppose G contains an element c of infinite order. We may assume that $c \in Z$. Since $\{c\} \cap G_0 = e$, we would have $\varphi\{c\} = O$. Since $c \in Z$, $c\,a$ would also be of infinite order, so that $\varphi\{a\,c\} = O$. This is a contradiction, because

$$\varphi\{a\} \subseteq \varphi\{a\,c, c\} = \varphi\{a\,c\} \cup \varphi\{c\} = 0.$$

Hence all elements of G have finite order; in other words G is a torsion group. Take any element b of G with order a power of q ($q \neq p$). Then the subgroup $H = \{a, b\}$ is a finite group such that $\varphi(H) = \varphi(G_0) = I$. Any maximal subgroup U of H, not containing G_0, is mapped on the maximal element of C, and hence there is only one such maximal subgroup of H. Hence H is a direct product of G_0 and $\{b\}$. In particular b is an element of Z. Let K be a subgroup of G, generated by all elements with order relatively prime to p. K is a part of E and a normal subgroup of G, and G/K is a p-group, which implies (4). Let u be an element of order p^α. The group $V = \{a, u\}$ is a finite p-group which is mapped on I. Hence every maximal subgroup of V contains G_0 if $V \supset G_0$. This implies in particular that V is cyclic. Hence either $G_0 \supseteq \{u\}$ or $\{u\} \supseteq G_0$.

This proves (3) and moreover we conclude that $E = (G_0 \wedge E) \vee K$, which implies $E \subseteq Z$ as stated in (5).

Assume conversely that a torsion group G satisfies the properties (1) to (5). Let $E_0 = E \wedge G_0$ and define a mapping φ by

$$\varphi(U) = (U \vee E_0) \wedge G_0, \text{ for every subgroup } U \text{ of } G.$$

Then it is easy to verify that φ is a complete L-homomorphism of G onto $L(G_0/E_0)$. Since $L(G_0/E_0)$ is a finite chain of dimension not less than n, G admits a complete L-homomorphism onto a finite chain of dimension n.

Remark. If G is *locally finite*, i. e. if a finite subset of G (consisting of elements of finite order) generates a finite subgroup of G, then the totality of elements with orders relatively prime to p is a normal subgroup K of G and E is the direct product of K and $E \wedge G_0$. Moreover the factor group G/K is either a cyclic group of finite order, a group of type p^∞ in the sense of PRÜFER, or a generalized quaternion group. The first statement follows from the splitting theorem of SCHUR, and the second one follows from the fact: G/K contains only one subgroup of order p.

The following theorem is a consequence of (3.1) (WHITMAN [*2*], ZAPPA [*6*] and SUZUKI [*2*]).

Theorem 3. *A group G admits a complete L-homomorphism φ onto a finite cyclic group Z of order $\prod_{i=1}^n q_i^{e_i}$ if and only if G is a torsion group and there exist distinct prime numbers p_i and normal subgroups G_i and E_i $(i = 1, 2, \ldots, n)$ with the following properties:*

(1) *G_i is a cyclic group of order $p_i^{f_i}$,*

(2) $(G_i : G_i \wedge E_i) \geq p_i^{e_i}$,

(3) *every p_i-subgroup of G either contains or is contained in G_i,*

(4) *each G/E_i is a p_i-group $(i = 1, 2, \ldots, n)$, and*

(5) *every element of G_i commutes with every element of E_i.*

Proof. The subgroup lattice of Z is a direct product of chains C_i with dimension e_i. Hence there are complete L-homomorphisms θ_i of $L(Z)$ onto C_i. The combined mapping $\theta_i \varphi$ is a complete L-homomorphism of G onto C_i. Let G_i and E_i be the u-kernel and the l-kernel of $\theta_i \varphi$ respectively. Let $(G_i : e) = p_i^{f_i}$. Then the set of primes $p_1, \ldots, p_n$ and G_i and E_i satisfy the requirements of this theorem.

Assume conversely that G has the structure of theorem 3. G admits by (3.1) a complete L-homomorphism φ_i onto a finite chain C_i of dimension e_i. The mapping φ of $L(G)$ defined by $\varphi(U) = (\varphi_1(U), \ldots, \varphi_n(U))$ is proved to be a complete L-homomorphism of G onto $\prod C_i \cong L(Z)$.

Theorem 4. *A group G admits a complete L-homomorphism onto an infinite cyclic group if and only if G is a torsion-free abelian group of rank* 1.

Proof. Suppose G is a torsion-free abelian group of rank 1. Any element $u \neq 1$ of G generates an infinite cyclic subgroup Z. The mapping φ defined by $\varphi(U) = U \cap Z$ maps $L(G)$ onto $L(Z)$. It preserves intersections, i. e.

$$\varphi(\cap_\lambda U_\lambda) = \cap_\lambda \varphi(U_\lambda) \text{ for any index set } \Lambda.$$

At the same time it preserves unions. It is clear that

$$\varphi(\cup_\lambda U_\lambda) = (\cup_\lambda U_\lambda) \cap Z \subseteq \cup_\lambda (U_\lambda \cap Z).$$

Take an element $z \in (\cup_\lambda U_\lambda) \cap Z$. Then there is a finite set of λ's, say $\lambda_1, \ldots, \lambda_n$ such that $z \in \cup_{i=1}^n U_{\lambda_i}$. Let $V = \cup_\lambda (U_\lambda \cap Z)$. Then $V \neq e$. Since Z is infinite cyclic there is a finite set $\mu_1, \ldots, \mu_m$ such that $V = \cup_{j=1}^m (U_{\mu_j} \cap Z)$. Since G is a D-group by theorem I. 2, p. 4,

$$\begin{aligned} V &= \cup_{j=1}^m (U_{\mu_j} \cap Z) \cup \cup_{i=1}^n (U_{\lambda_i} \cap Z) \\ &= \big((\cup_{j=1}^m U_{\mu_j}) \cap Z\big) \cup \big((\cup_{i=1}^n U_{\lambda_i}) \cap Z\big) \\ &= \big((\cup_{j=1}^m U_{\mu_j}) \cup (\cup_{i=1}^n U_{\lambda_i})\big) \cap Z \ni z. \end{aligned}$$

This shows that $V \supseteq (\cup_\lambda U_\lambda) \cap Z$, and φ defines a complete L-homomorphism of G onto $L(Z)$.

Suppose conversely that G admits a complete L-homomorphism φ onto an infinite cyclic group Z. Let G_0 be the u-kernel, and E be the l-kernel of φ. First of all we shall prove that G_0 is an infinite cyclic group and $E = e$.

Take any proper subgroup U of Z. Then as proved above the mapping $\varphi_U : V \to V \cap U$ defines a complete L-homomorphism of Z onto $L(U)$. Hence $\varphi_U \varphi$ is a complete L-homomorphism of G onto $L(U)$. The u-kernel of $\varphi_U \varphi$ is a minimal subgroup U^* of G_0 such that $\varphi(U^*) = U$. U^* is a normal subgroup of G by theorem 1, p. 58, and φ induces a complete L-homomorphism $\overline{\varphi}$ of G_0/U^* onto a finite cyclic group Z/U. By definition the u-kernel of $\overline{\varphi}$ coincides with G_0/U^*. Hence by theorem 3, G_0/U^* is a finite cyclic group. Let $E_0 = E \cap G_0$, and consider a subgroup H of G_0 such that $H \supset E_0$. Then $\varphi(H) = U$ is a proper subgroup of Z, and so H contains U^*. Hence H is a normal subgroup of G_0 with the cyclic factor group G_0/H. Since Z is an infinite cyclic group and since φ is complete, we take a series of subgroups H_i such that $H_i \supseteq H_{i+1}$, $\cap_i H_i = E_0$ and none of the H_i is equal to E_0. Then each H_i is normal in G_0 and the factor group G_0/H_i is finite and cyclic. Hence the factor group G_0/E_0 is abelian, and hence infinite cyclic. We may therefore take an element u of G_0 such that $G_0 = \{u, E_0\}$.

Applying φ, we conclude that

$$\varphi\{u\} = \varphi\{u, E_0\} = \varphi(G_0) = Z.$$

By definition of G_0, this shows that $\{u\} \supseteq G_0$, or $G_0 = \{u\}$. Since $\{u\} \cap E_0 = e$, this implies that $E_0 = e$. Hence every element v of E commutes with u, and so $u\,v$ is also of infinite order. Now we have

$$\varphi\{u\,v\} = \varphi\{u\,v, v\} = \varphi\{u, v\} = \varphi\{u\},$$

so that $\{u\,v\} \supseteq \{u\}$ since $\{u\} = G_0$. But this is possible only when $v = 1$. Hence E consists only of the identity element.

Take any maximal subgroup V_1 of G_0, and let $p = (G_0 : V_1)$. $\varphi(V_1) = Z_1$ is a maximal subgroup of Z, and Z contains a chain of subgroups

$$Z_1 \supset Z_2 \supset \cdots \supset Z_t \supset \cdots$$

such that $\bigcap_i Z_i = e$ and $(Z_i : Z_{i+1}) = (Z : Z_1)$ for all $i = 1, 2, \ldots$. Since G_0 is infinite cyclic, we can take the maximal subgroup V_t of G_0 with the property $\varphi(V_t) = Z_t$. φ induces now a complete L-homomorphism of G/V_t onto a finite chain $L(Z/Z_t)$. By (3.1) there is a normal subgroup N_t such that G/N_t is a p-group, and $\varphi(N_t) = Z_t$. Since we selected V_t as maximal among those subgroups of G_0 which are mapped upon Z_t we must have $N_t \cap G_0 = V_t$, and N_t is a maximal subgroup which is mapped upon Z_t. Hence $N_t \supseteq N_{t+1}$ and $N_1 = N_t \cup V_1$ for all t. Hence N_1/N_t is a finite cyclic group. Applying φ, we see that

$$\varphi(\bigcap_i N_i) = \bigcap_i \varphi(N_i) = \bigcap_i Z_i = e.$$

Since the l-kernel of φ consists of the identity element only, this shows that $\bigcap_i N_i = e$, and hence N_1 is abelian.

Now we get a normal subgroup $N(p) = N_1$ such that $N(p)$ is abelian and the factor group $G/N(p)$ is a p-group. If we take another maximal subgroup M of G_0 such that $(G_0 : M) = q \neq p$, the same consideration as above shows the existence of an abelian normal subgroup $N(q)$ of G such that the factor group $G/N(q)$ is a q-group. Since $p \neq q$, we have $G = N_t \cup N(q)$. Hence each factor group G/N_t is isomorphic with $N(q)/N(q) \cap N_t$, which is abelian. Since $\cap N_t = e$, G must be abelian. If we take an element w of G such that $\{w\} \cap G_0 = e$, then we have

$$\varphi\{w\} = \varphi(\{w\} \cap G_0) = e.$$

But the l-kernel of φ is e, so that $w = 1$. Hence G is a torsion-free abelian group of rank 1.

3. General properties of complete L-homomorphisms.

Let φ be a complete L-homomorphism of a group G onto a lattice L. Denote by G_0 the u-kernel, and by E the l-kernel of φ. If $E = G$, φ is trivial. We term φ *proper*, if φ is neither an isomorphism nor trivial.

Theorem 5. *Let φ be a non-trivial complete L-homomorphism of G. The l-kernel E of φ consists of elements of finite order, and if G contains an element of infinite order, $E = e$.*

Proof. By assumption there exists a cyclic subgroup Z generated by $u: \{u\} = Z$ such that $\varphi(Z) = a \neq O$. The interval a/O is isomorphic with the subgroup lattice of some cyclic group. Now φ induces a complete, non-trivial L-homomorphism $\bar{\varphi}$ of $Z \cup E$ onto a/O. Hence theorems 3 and 4 show that E consists of elements of finite order. Moreover if $E \neq e$, every element outside of E has finite order, so that G has no element of infinite order.

We consider the case $E \neq e$. Then G is a torsion group. Let $E_0 = E \cap G_0$, $G_1 = G_0 E$, and denote by Π the set of all primes which occur as orders of elements in G_1/E_0.

Suppose that there is an element of order p in the factor group G/G_1. Then we may take an element u of G such that the order of u is p^α $(\alpha \geq 1)$, and $u \notin G_1$. Since $u \notin G_1$, we have $\varphi(Z) \neq O$, where $Z = \{u\}$. Hence

$$\varphi(Z \cap G_0) = \varphi(Z) \cap \varphi(G_0) = \varphi(Z) \neq O.$$

This shows that $Z \cap G_0$ is not contained in E. Hence $p \in \Pi$. Π is therefore the set of primes which occur as orders of elements of G/E.

Take $p \in \Pi$. Let N_p be the totality of elements in E, whose orders are powers of p. By definition of Π there is an element u of G with order p^α, not contained in E. Let $Z = \{u\}$. Then $\varphi(Z) \neq O$, and φ induces a proper L-homomorphism $\bar{\varphi}$ of $Z \cup E$ onto $\varphi(Z)/O$. Since $\varphi(Z)/O$ is a finite chain, we may apply (3.1) to $\bar{\varphi}$. From the property (3) of (3.1) of p. 59, we conclude that $N_p \subseteq Z$ and from (5) we see that N_p is in the center of E. Hence N_p is a finite cyclic subgroup of E, which is contained in the center of E, and every p-subgroup of G either contains or is contained in N_p.

N_p is actually in the center of G_1. Let now Z be the centralizer of N_p in G_0. If $Z = G_0$. The proof is complete. Suppose $Z \neq G_0$. Then Z would be a normal subgroup of G_0 such that G_0/Z is a cyclic group of finite order n and $(n, p) = 1$. Take an element v of G_0 such that $G_0 = \{v, Z\}$. $H = \{v\}$ and N_p generate a finite subgroup $U = H \cup N_p$ and φ induces an L-homomorphism $\tilde{\varphi}$ of U onto $\varphi\{v\}/O$. Then the u-kernel V of $\tilde{\varphi}$ is normal and is contained in H. Hence $V \subseteq Z$, a contradiction, since $\varphi\{v\} = \tilde{\varphi}(V) \subseteq \varphi(Z)$, or $\varphi(G_0) = \varphi(Z)$.

Set $N = \bigcup_{p \in \Pi} N_p$. N is a direct product of cyclic groups N_p, and is contained in the center of G_1. Since each N_p is a subgroup of G_0, N is contained in the intersection E_0 of E and G_0. Hence the order of any element of E/E_0 is relatively prime to each $p \in \Pi$.

We may summarize these results as follows.

Proposition 3.2. *Let φ be a non-trivial complete L-homomorphism of G onto a lattice L, G_0 be its u-kernel, and E be its l-kernel. If $E \neq e$, then G is a torsion group. Define Π as the set of all primes which occur as orders of elements in $G_0/G_0 \cap E$.*

(1) *If p is the order of some element of $G/G_0 \cup E$, then $p \in \Pi$,*

(2) *the orders of elements of $E/E \cap G_0$ are relatively prime to every $p \in \Pi$,*

(3) *the totality of elements of E with p-power order forms a finite cyclic group N_p such that N_p is contained in the center of $G_0 \cup E$, and every p-subgroup of G either contains or is contained in N_p.*

If the group G is locally finite, in particular if G is a finite group, then we may apply the splitting theorem of Schur to obtain better results. It implies that the extension of E over N splits, so that E is a direct product of N and another subgroup K, which consists of elements of E whose orders are relatively prime to all $p \in \Pi$. K is a normal subgroup. Hence $K \cap G_0$ is normal in G_0 and the extension over $K \cap G_0$ again splits. This means that $G_0 \cap K = e$ and $E \cap G_0 = N$. The same is true, if we assume G to be solvable.

4. L-homomorphisms induced by group-homomorphisms.

Let G be any group, and H be a subgroup of G. The mapping φ of $L(G)$ defined by $\varphi(U) = U \cup H$ maps $L(G)$ onto the interval G/H, and preserves unions. Under what conditions will it preserve intersections? This problem has been solved by Zappa [7] for finite groups, by D. G. Higman [1] and Sato [4] for the general case.

Suppose that $\varphi: U \to U \cup H$ preserves finite intersections. Then φ defines an L-homomorphism of G onto the interval G/H, and H coincides with the l-kernel of φ. Hence by theorem 2 H is a normal subgroup of G. We assume in the following that φ is proper, i. e. $H \neq G, e$. First of all we prove that G/H contains no element of infinite order. By way of contradiction assume that G/H contains an element X of infinite order. We may take a representative x of X modulo H and let $U = \{x\}$. Then φ defines an L-homomorphism $\bar{\varphi}$ of $U \cup H$ onto the subgroup lattice of $L(U\,H/H)$. Since H is normal, for any proper subgroup V of U, $V \cup H$ is a proper subgroup of $U \cup H$. This means that U is the u-kernel of $\bar{\varphi}$ and hence is a normal subgroup of $U \cup H$. Hence $\bar{\varphi}$ is a complete L-homomorphism of $U \cup H$ with H as its l-kernel. This contradicts theorem 5, since we have assumed that $H \neq e$. Hence all elements of G/H have finite order. We denote by Π the set of all primes which occur as orders of elements of G/H.

Suppose now that G contains an element u of finite order, which is not contained in H. Set $U = \{u\}$. Again φ induces an L-homomorphism $\bar{\varphi}$ of $U \cup H$ onto $L(U\,H/H)$, and $\bar{\varphi}$ is complete. By theorem 3, all elements of H have finite order. Hence all elements of G have finite order. Consider an element v of G such that the order of v is p^α ($p \in \Pi$, $\alpha > 0$) and $v \notin H$. $Z = \{v\}$ is the u-kernel of the L-homomorphism $\tilde{\varphi}$ of $Z \cup H$ onto $L(Z\,H/H)$, induced by φ. That $\tilde{\varphi}$ is complete may be verified. Hence, by (3.1), v commutes with every element of H and $Z \cap H$ coincides with the totality of elements of H with p-power order. We have thus proved the necessity of the following conditions (1) and (2) (SATO [*4*], HIGMAN [*1*] and ZAPPA [*7*]).

Theorem 6. *Let G be a torsion group, and let H be a subgroup of G. The mapping*

$$\varphi : \varphi(U) = U \cup H \text{ for } U \subseteq G,$$

defines an L-homomorphism of G, if and only if

(1) *H is a normal subgroup of G, and*

(2) *if Π is the totality of prime numbers which occur as orders of elements of G/H, and if N_p is the totality of elements of H with p power orders, then for every $p \in \Pi$,*

(2a) *any element u of order a power of p in G commutes with all elements of H, and*

(2b) *we have either $\{u\} \supseteq N_p$, or $N_p \supseteq \{u\}$.*

Proof of sufficiency. Assume now that G and H satisfy these conditions. It is sufficient to prove that $(\cap_\lambda U_\lambda) \cup H$ contains $\cap_\lambda (U_\lambda \cup H)$ for any set of subgroups U_λ ($\lambda \in \Lambda$). Take an element t of prime power order, say of order p^α, in $\cap_\lambda (U_\lambda \cup H)$. We may assume $t \notin H$. Then $p \in \Pi$. By (2a) of this theorem t commutes with all elements of H. Any element $h \in H$ is a product of the form $n\,z$ where the order of n is a power of p, and that of z is prime to p. By (2b) n is a power of t; $n = t^a$, and $a \equiv 0 \pmod p$. Hence t is contained in $\{n\,t\}$ and $\{n\,t\} \subseteq \{t\,h\}$. Now $t \in U_\lambda \cup H$ implies that there is an element $n_\lambda \in H$ such that $t\,n_\lambda \in U_\lambda$. As proved above t is in $\{t\,n_\lambda\}$ and hence $t \in \cap_\lambda U_\lambda$. Hence $(\cap_\lambda U_\lambda) \cup H \supseteq \cap_\lambda (U_\lambda \cup H)$ as was to be shown.

Corollary. *If, under the same assumptions as those of theorem* 6, *the mapping $U \to U \cup H$ defines an L-homomorphism, then it is complete.*

If the group G is *locally finite*, we can prove more. The totality of elements of H whose orders are relatively prime to all $p \in \Pi$ forms a normal subgroup K of G. Since φ is complete, there is a smallest subgroup T such that $T \cup H = G$. This T is a normal subgroup, since it is the u-kernel of φ. Hence G is a direct product of T and K, and $T \cap H$ is in the center of G (ZAPPA [*7*] and D. G. HIGMAN [*1*]).

If G contains an element of infinite order, the situation is rather complicated. We have the following theorem (SATO [4], p. 232, theorem 2).

Theorem 7. *Let G be a group containing an element of infinite order, and let H be a subgroup of G. The mapping*

$$\varphi : \varphi(U) = U \cup H \text{ for } U \subseteq G$$

is an L-homomorphism of G, if and only if

(1) *H is a normal subgroup of G,*

(2) *the factor group G/H is a torsion group,*

(3) *all elements of finite order of G are contained in H, and their orders are relatively prime to these orders of elements in G/H,*

(4) *if m and n are elements of infinite order in H, then*

$$\{m\} \cap \{n\} \neq e, \text{ and}$$

(5) *if X is any coset modulo H, and if x is any element in X, then for any $h \in H$, there exists an integer k such that*

$$x^k \in X, \text{ and } x^k h = h x^k.$$

Proof. Assume that the mapping φ preserves intersections. Conditions (1) and (2) have been proved before. If an element u of G is not contained in H, then u is of infinite order, since we have assumed that G is not a torsion group. Take any element n of H, then $\{u, H\} = \{u\, n, H\}$. Hence $(\{u\} \cap \{u\, n\}) \cup H = \{u, H\}$. This means that $\{u\} \cap \{u\, n\} \neq e$. If $\{u\} \cap \{u\, n\} = \{v\}$, then $v = u^k$, v commutes with n, and moreover $\{u, H\} = \{v, H\}$. If n is of finite order, φ induces an L-homomorphism of $\{v, n\}/\{v\} \cap H$ onto $L(\{v, H\}/H)$. Hence by (3.2) of p. 64 the order of n is relatively prime to any prime divisor of the order of elements in G/H. If n is of infinite order, $\{v\} \cap \{n\} \neq e$, since $\{v\} \cap \{v\, n\} \neq e$. Hence if m and n are elements of H with infinite order, then $\{m\} \cap \{n\} \neq e$.

Assume conversely that G and H satisfy all conditions (1) to (5). Take any coset X mod H, and two elements u and v of G in X. We want to show that $\{u\} \cap \{v\} \cap X$ is not empty. Since $u, v \in X$, $v = u\, h$ where $h \in H$. Now by (5) there exists a power of u, $x = u^k$ such that $x \in X$ and $x\, h = h\, x$. Hence we have $x\, v = v\, x$, $v = x\, w$, $w \in H$. Let n be the order of X in G/H. Then x^n and v^n are in H, and so $\{x\} \cap \{v\} \neq e$ by (4). We may take the smallest integer α such that $x^\alpha = v^\beta$. We want to show $(\alpha, n) = 1$. If this is done, the proof will be completed. Since $x^\alpha = v^\beta = (x\, w)^\beta = x^\beta w^\beta$, we have $x^{\alpha - \beta} = w^\beta \in H$, or $\alpha \equiv \beta \pmod{n}$. If there were some prime p such that $p \mid \alpha$ and $p \mid n$, p would also divide β. Hence the p-th power of $x^{\alpha/p} v^{-\beta/p}$ would be the identity element, but we assumed that every element of finite order has order relatively prime to p. Hence we would have $x^{\alpha/p} = v^{\beta/p}$, which contradicts the minimality of α. Now consider $X \in (U \cup H) \cap (V \cup H)$ for any subgroups U and V of G. We may take $u \in X \cap U$ and $v \in X \cap V$. Since $\{u\} \cap \{v\} \cap X$ is not empty as proved above, we may take $t \in \{u\} \cap \{v\} \cap X$. Since $t \in \{u\} \subseteq U$, and $t \in \{v\} \subseteq V, t \in U \cap V$, or $X \in (U \cap V) \cup H$. This proves $(U \cap V) \cup H = (U \cup H) \cap (V \cup H)$, and the mapping $U \to U \cup H$ preserves intersections.

Let σ be a homomorphism of G onto another group G'. σ induces a lattice mapping φ of $L(G)$ onto $L(G')$. If we denote by N the kernel of this homomorphism σ, G' may be identified with G/N and σ with the natural homomorphism of G onto G/N. By this identification, φ may be regarded as the lattice mapping $U \to U \cup B$. Hence theorems 6 and 7 give conditions under which σ induces an L-homomorphism of G.

If σ induces a complete L-homomorphism, all elements of G have finite order by theorem 5, and we do not have the case of theorem 7.

In particular suppose that σ is an endomorphism of G, and σ induces an L-homomorphism, which maps every subgroup U of G upon the intersection of U and $\sigma(G)$. This case has been treated by Zappa [7] and D. G. Higman [1].

Theorem 8. *Let H be a proper subgroup of G. Then there exists an endomorphism σ of G such that $\sigma(U) = U \cap H$ for every subgroup U of G if and only if every element in G has finite order, H is a direct factor and the orders of the elements in H are relatively prime to the orders of the elements of G/H.*

Proof. Suppose that the endomorphism σ is such that $\sigma(U) = U \cap H$. Then σ induces a complete L-homomorphism φ and H is the u-kernel, and N, the kernel of σ, is the l-kernel of φ. Since $N \cap H = e$ and $NH = G$, the necessity of the condition of this theorem follows now from (3.2).

Assume conversely that G has the properties of this theorem. Then $G = N \times H$ for some group N, and $L(G) = L(N) \times L(H)$ by theorem I. 4. Every subgroup U of G has the form $(U \cap N) \times (U \cap H)$. Hence the decomposition homomorphism of G onto H induces an L-homomorphism of G onto H.

We point out a special case of theorem 6. Let G be a group, all of whose elements have finite order, and Z be a cyclic p-subgroup of G such that Z is contained in the center of G and every p-subgroup of G either contains or is contained in Z. Then the mapping of G defined by $U \to U \cup Z$ for any subgroup U of G defines a complete L-homomorphism of G onto $L(G/Z)$. If G is a finite group, the last condition can be expressed as follows: Z is a central p-subgroup of G and p-Sylow subgroups of G are either cyclic or generalized quaternion groups.

5. Incomplete L-homomorphisms.

If an L-homomorphism φ is not complete most of the results in sections 2 and 3 cease to be true. SATO [5] studied general L-homomorphisms on a chain.

Let φ be an L-homomorphism of a group G onto a chain C and let E be the l-kernel of φ. It is possible that $E = G$. If so, φ maps clearly the l-kernel E upon the greatest element I of C. On the other hand, if $E \neq G$, then φ maps E upon the least element O of C. SATO called an L-homomorphism φ *regular* if φ maps its l-kernel upon O and proved the following result.

Proposition 3.3 (SATO [5]). *There exists a regular incomplete L-homomorphism of a group G onto the 1-dimensional chain, if and only if G contains proper normal subgroup E such that E contains all elements of finite order in G and either E is a torsion group or else G/E is a torsion group containing only one minimal subgroup whose order is not a divisor of the order of any element of G, and we have the following two properties:*

$$\{a\} \cap Z(b) \nsubseteq E \text{ and } \{a\} \cap \{c\} \neq e$$

where $a \nsubseteq E$, $b \in E$, $Z(b)$ is the centralizer of b and c is any element of G with infinite order.

Proof. Suppose φ is a regular incomplete L-homomorphism of G onto the one dimensional chain C. The l-kernel E of φ is a proper normal subgroup. If there is an element of finite order outside E, then there is the smallest subgroup of G which is mapped upon I. Hence φ is complete. We conclude therefore that every element outside E has infinite order.

If we take two elements a und c outside E, then

$$\varphi(\{a\} \cap \{c\}) = \varphi\{a\} \cap \varphi\{c\} = I.$$

Hence $\{a\} \cap \{c\} \nsubseteq E$. If $b \in E$, then $ab \nsubseteq E$ and therefore $\{a\} \cap \{ab\} \nsubseteq E$. This implies that a power of a commutes with ab and hence with b. Thus we have shown

$$\{a\} \cap Z(b) \nsubseteq E \text{ if } a \notin E,\ b \in E.$$

We shall distinguish two cases. The first case is that there is an element a such that $\{a\} \cap E = e$. We shall show that E is a torsion group. Let $b \in E$. Since $\{a\} \cap Z(b) \neq e$, a power a^k of a commutes with b. Now $\{a^k\} \cap \{a^k b\} \neq e$. Hence there are integers m and n such that

$$a^{km} = (a^k b)^n, \text{ so } b^n = a^{k(m-n)}.$$

Hence $b^n \in \{a\} \cap E = e$. This proves that E is a torsion group.

Now we consider the second case that $\{a\} \cap E \neq e$ for all elements $a \neq 1$ of G. In this case the factor group G/E must be a torsion group.

If U is a smallest subgroup of G such that $U \supset E$, then $\varphi(U) = I$. Hence G/E can not contain two different minimal subgroups. Let $p = (U:E)$. Let us take an element a of U, not in E. If $b \in E$, then some power a^k is not in E and commutes with b. Since $\{a^k\} \cap \{a^k b\} \nsubseteq E$, there are integres m and n such that

$$a^{km} = (a^k b)^n \notin E.$$

If b is of finite order, the above relation implies that $b^n = 1$ and $a^{kn} \notin E$. Since $p = (U:E)$, $k\,n$ is not a multiple of p. Hence p is not a divisor of orders of any elements of E. If the order of b is infinite then we conclude that $\{a\} \cap \{b\} \neq e$. The necessity of conditions of this proposition is thus proved.

Conversely assume a group G has the structure of (3.3). We define

$$\varphi(V) = \begin{cases} I \text{ if } V \nsubseteq E \\ O \text{ if } V \subseteq E \end{cases}$$

In order to prove that φ is an L-homomorphism it suffices to show

$$V \cap W \nsubseteq E \text{ whenever } V \nsubseteq E,\ W \nsubseteq E.$$

If E is a torsion group V and W contain elements of infinite order. Hence it follows from the last condition that $V \cap W$ contains also an element of infinite order and a fortiori $V \cap W$ is not a part of E. Assume now that G/E is a torsion group. Then there are elements a and c outside E such that $a \in V$, $c \in W$ and $a = b\,c$ where $b \in E$. By the conditions of (3.3) there is a power $c^k \notin E$ of c which commutes with b. Then $a^k = b'\,c^k$ and $b' \in E$ commutes with c^k. If b' is of finite order n, n is prime to the order of a in G/E. Hence $a^{kn} = c^{kn} \notin E$, and hence $V \cap W \nsubseteq E$. If b' is of infinite order, let s be the smallest positive integer such that $b'^s = c^{kt}$. The existence of such s follows from the last condition. From the minimality of s it follows that s is not a multiple of the prime divisor p of the order of a in G/E, since otherwise $b'^{s'} = c^{kt}$ where $s = s'\,p$ and $t = t'\,p$. Now

$$a^{ks} = b'^s\,c^{ks} = c^{k(s+t)}$$

is not an element of E and hence $V \cap W \nsubseteq E$. Thus the assertion is completely proved.

A regular L-homomorphism φ of a group upon a one dimensional chain satisfies the equation

$$\varphi(\cup_{\lambda \in \Lambda} U_\lambda) = \cup_{\lambda \in \Lambda} \varphi(U_\lambda)$$

for any number of subgroups U_λ $(\lambda \in \Lambda)$. Such an L-homomorphism may be called (upper) *semi-complete.* Using (3.3) one may obtain results on upper semi-complete L-homomorphisms onto a (finite) chain or a direct product of chains (cf. SATO [5]). If H is a subgroup of G the mapping ψ defined by $\psi(U) = U \cup H$ satisfies the above equation. Hence ψ is upper semi-complete if ψ preserves finite intersections. The lower semi-completeness is defined as dual of the above concept. No result corresponding to (3.3) is known for lower semi-complete L-homomorphisms.

6. L-homomorphisms of finite groups.

The remaining part of this chapter will be devoted to the study of L-homomorphisms of *finite* groups. Every L-homomorphism of a finite group is complete, and so we may apply the results of the preceding sections. For finite groups, we may make use of somewhat deeper theorems on finite groups, in particular theorems of SYLOW, to get better results. This was done by ZAPPA [*6, 7, 8* and *11*] and SUZUKI [*2*].

Proposition 3.4. *A p-group G admits a proper L-homomorphism φ if and only if G is either a cyclic group or a generalized quaternion group.*

Proof. If the u-kernel G_0 of φ differs from G, the subgroup H covering G_0 is cyclic, since it contains only one maximal subgroup. Hence (3.4) follows from (3.1) on p. 59 in this first case.

Assume next that G_0 coincides with G. Since φ is proper, φ induces a proper L-homomorphism into every maximal subgroup of G. Using an inductive argument, every maximal subgroup of G contains only one subgroup of order p. If G contained two different minimal subgroups, G would be an abelian group of order p^2, and would therefore admit no proper L-homomorphism. This proves (3.4).

Proposition 3.5. *If φ induces a proper L-homomorphism of a cyclic p-SYLOW subgroup S of G, then G contains a p-SYLOW complement which is normal.*

Proof. Using induction on the order of G, we want to show that S is in the center of its normalizer. If this is done (3.5) follows from a theorem of BURNSIDE (Cf. ZASSENHAUS [*1*], p. 133). We may assume that S is normal with a cyclic p-SYLOW complement K. Put $S = \{a\}$ and $K = \{b\}$, then we have $b\,a\,b^{-1} = a^r$. If $r \not\equiv 1$ (modulo the order of a), G would not admit any proper L-homomorphism contradicting our assumption. Hence $S \cup K = S \times K$.

We can improve (3.2) on p. 64 in case of finite groups (SUZUKI [*2*]).

Proposition 3.6. *Let φ be an L-homomorphism of a finite group G. Denote by G_0 and E the u- and l-kernel of φ respectively. Then G contains a normal subgroup N and a subgroup H such that*

(1) *$G = N\,H$ and $N \cap H = e$,*

(2) *the orders of N and H are relatively prime,*

(3) *H contains the u-kernel G_0,*

(4) *the l-kernel E is the direct product of N and $E_0 = E \cap G_0$,*

(5) *E_0 is a cyclic group, contained in the center of G, and*

(6) *if a prime p divides the order of E_0 or of H/G_0, then it divides the order of G_0/E_0.*

Proof. By (3.2), G contains a normal subgroup N such that the order of any element in N is relatively prime to all primes which occur as orders of elements in G/E. Hence N is a normal subgroup, whose order is relatively prime to the index $(G:N)$. In virtue of SCHUR's theorem G contains a subgroup H with the properties (1) and (2). Hence applying φ, we see that

$$\varphi(G) = \varphi(N) \cup \varphi(H) = \varphi(H).$$

Therefore H contains the u-kernel G_0. Hence we have, in particular, $G_0 \cap N = e$, which implies (4).

The property (5) is proved as follows. By (3.2) E_0 is a direct product of cyclic p-groups N_p, and so E_0 itself is cyclic. Suppose $N_p \neq e$. Take a p-SYLOW subgroup S of G. Then S contains N_p and is cyclic or a generalized quaternion group by (3.4). If S is a generalized quaternion group, N_p must be of order 2 and hence is contained in the center of G. We assume that S is cyclic. By (3.5), a p-SYLOW complement K in G is normal. As N_p is normal, we have $K \cup N_p = K \times N_p$. Hence N_p is in the center of G.

Now if a prime p divides the order of E_0, it is a prime factor of the order of G_0/E_0, since all maximal subgroups of G_0 contain E_0. If it divides the order of H/G_0, φ does not induce a projectivity into a p-SYLOW subgroup S of H. If $S \cap G_0 = e$, we have $\varphi(S) = \varphi(S) \cap \varphi(G_0) = O$, which implies that $S \subseteq E$ or $S \subseteq H \cap E = E_0$. This is impossible. Hence $S \cap G_0 \neq e$ and p divides the order of G_0/E_0.

To determine the structure of H/G_0 in the above notation, we need

Proposition 3.7. *If a generalized quaternion group Q is mapped by φ onto a chain of dimension two, H is the direct product of its 2-SYLOW subgroup S_2 and its 2-SYLOW complement K and L is a direct product of $\varphi(S_2)$ and $\varphi(K)$.*

Proof. First, using induction on the order of G, we prove that G has a normal 2-SYLOW complement. By (3.4) the 2-SYLOW subgroups of G are generalized quaternion groups. Take any proper subgroup V of G. If its 2-SYLOW subgroup is cyclic, V has a normal 2-SYLOW complement by a theorem of BURNSIDE (Cf. ZASSENHAUS [1], p. 133). If its 2-SYLOW subgroup is a generalized quaternion group, then the existence of a normal 2-SYLOW complement follows from the inductive hypothesis. Hence every subgroup of G has a normal 2-Sylow complement. By a theorem of Itô [1], G has also a normal 2-Sylow complement, or all proper subgroups of G are nilpotent. In the latter case, if its 2-Sylow complement were not normal, G would be of order $p^\alpha 2^\beta$ (p is a prime greater than 2). The structure of such a group has been completely determined by Schmidt and Iwasawa. We can prove by direct examination that our assumption does not hold in this case. Hence G has a normal 2-Sylow complement.

Next using again induction on the order of H, we prove that H is a direct product of its 2-Sylow subgroup and the 2-Sylow complement. We shall denote by K the 2-Sylow complement of H and assume for a while that the l-kernel of φ coincides with e. Considering normalizers of Sylow subgroups of K, we can assume K to be a p-group ($p > 2$). If K is cyclic, the centralizer of K contains the center Z of a 2-Sylow subgroup S_2. Since

$$\varphi(K \cup Z) = \varphi(K) \cup \varphi(Z) = \varphi(K) \cup \varphi(S_2) = \varphi(H),$$

$K Z$ contains the u-kernel of φ and is the direct product of K and Z. Hence we have

$$L = (\varphi(K)/O) \times (\varphi(Z)/O).$$

Let ψ be the natural homomorphism from L onto $\varphi(K)/O$. Then $\psi\varphi$ is an L-homomorphism from H onto $\varphi(K)/O$ and S_2 is the l-kernel of $\psi\varphi$, since we assumed the l-kernel of φ to coincide with e. Hence by theorem 2, S_2 is normal in H and we have $H = K \times S_2$.

If K is not cyclic, φ induces an L-isomorphism from K into L by (3.4). We can, therefore, assume that S_2 is maximal. If the center Z of S_2 is normal in H, S_2 is normal in the same way as above. If Z were not normal in H, Z would be conjugate to another group Z_1. Z_1 would be the center of a 2-Sylow subgroup Q and $Q \neq S_2$. Then we would have

$$\varphi(Z \cup Z_1) = \varphi(Z) \cup \varphi(Z_1) = \varphi(S_2) \cup \varphi(Q) = \varphi(H)$$

and hence $Z \cup Z_1 \supseteq K$. This implies that K would be cyclic, which gives a contradiction.

If the l-kernel E_0 of φ in H differs from e, the l-kernel of φ in H/E_0 coincides with e. Hence the 2-Sylow subgroup $\overline{V}$ of H/E_0 is normal. Let V be a subgroup of H corresponding to $\overline{V}$ by the natural homomorphism from H onto H/E_0. Then V is normal in H. Take the normalizer N_2 of a 2-Sylow subgroup S_2 of H. Then we have $N_2 V = H$ because $S_2 \subseteq V$. On the other hand,

$$N_2 V = N_2 \cup S_2 \cup E_0 = N_2 \cup E_0.$$

Hence we have $N_2 E_0 = H$, which implies that $H = N_2$. Hence S_2 is normal and $H = S_2 \times K$.

Now we obtain the following proposition using the same notation as in (3.6), p. 70.

***Proposition** 3.8. The factor group H/G_0 is a nilpotent group, each of whose* SYLOW *subgroups is either cyclic or a dihedral group. If H/G_0 contains a dihedral group, H is the direct product of its* 2-SYLOW *subgroup and the* 2-SYLOW *complement. If, moreover, φ induces a proper L-homomorphism of $\overline{G}_0 = G_0/E_0$, $\overline{G}_0$ contains a normal subgroup $\overline{G}_1$ such that*

(1) *$\overline{G}_0/\overline{G}_1$ is cyclic,*

(2) *the order of $\overline{G}_0/\overline{G}_1$ is relatively prime to that of $\overline{G}_1$, and*

(3) *φ induces a projectivity of $\overline{G}_1$ into L.*

As special cases of these propositions we obtain the following theorems.

Theorem 9. *If none of the* SYLOW *complements of a group G is normal, any L-homomorphism of G onto a lattice L is either one of the natural homomorphisms from $L(G)$ onto direct components, or the L-homomorphism from G onto G/Z, where Z is the center of a* 2-SYLOW *subgroup, which is a generalized quaternion group, or a combination of these L-homomorphisms. In particular L is isomorphic to the subgroup lattice of some group.*

Since a projectivity maps a finite perfect group upon a finite perfect group (II, theorem 9, p. 46), we obtain the following result.

Theorem 10. *Let G be a perfect group. If φ is an L-homomorphism of G onto a subgroup lattice $L(H)$ of a group H, then H is perfect.*

7. The meet-homomorphisms.

A single-valued mapping φ of the subgroup lattice $L(G)$ of a group G onto a lattice L is a *meet-homomorphism,* if the relation

$$\varphi(U \cap V) = \varphi(U) \cap \varphi(V)$$

holds for every pair of subgroups U and V of G. The *join-homomorphism* is defined as the dual concept. (ZAPPA [8] studied these concepts and he termed them *lower* and *upper hemitropisms.*)

Let φ be a meet-homomorphism of G onto L. For every $a \in L$, we denote by $U(a)$ the smallest subgroup of G such that

$$\varphi\big(U(a)\big) = a.$$

The mapping of L into $L(G)$ defined by $a \to U(a)$ is a one-to-one mapping, satisfying

$$U(a) \supseteq U(b), \text{ if } a \geq b,$$

$$U(O) = e,$$

$$U(a \cap b) \supseteq U(a) \cap U(b),$$

and

$$U(a \cup b) = U(a) \cup U(b).$$

Proof of the last equation. If $\varphi(W) = a \cup b$, then

$$\varphi(W \cap U(a)) = \varphi(W) \cap \varphi(U(a)) = (a \cup b) \cap a = a.$$

Since $U(a)$ is the smallest subgroup such that $\varphi(U(a)) = a$, we have $W \cap U(a) \supseteq U(a)$, or $W \supseteq U(a)$. Similarly $W \supseteq U(b)$, and so

$$W \supseteq U(a) \cup U(b).$$

Hence

$$U(a \cup b) \supseteq U(a) \cup U(b).$$

On the other hand,

$$a \cup b = \varphi(W) \supseteq \varphi(U(a) \cup U(b)) \supseteq \varphi(U(a)) = a,$$

and similarly $\varphi\ (U(a) \cup U(b)) \supseteq b$. Hence

$$a \cup b \supseteq \varphi(U(a) \cup U(b)) \supseteq a \cup b,$$

or

$$\varphi(U(a) \cup U(b)) = a \cup b.$$

Hence

$$U(a) \cup U(b) = U(a \cup b).$$

Let Σ be a collection of subgroups of G satisfying

(1) $e \in \Sigma$,

(2) if U, $V \in \Sigma$, then $U \cup V \in \Sigma$.

Σ is then a finite set (since we consider only finite groups), and we define order in Σ by inclusion. By (1) and (2), Σ may be considered as a lattice. The join (meet) of U, $V \in \Sigma$ is the smallest (largest) subgroup W in Σ, which contains (is contained in) both U and V.

If H is any subgroup of G, H contains at least one member of Σ (for instance e by (1)). Define $\varphi(H)$ as the composition of all subgroups of H which belong to Σ. By (2) and the finiteness of G, $\varphi(H) \in \Sigma$. The mapping φ is a meet-homomorphism of G onto the lattice Σ. To prove this, it suffices to show

$$\varphi(H \cap K) = \varphi(H) \wedge \varphi(K)$$

for every pair of subgroups H and K of G. It is clear that

$$H \cap K \supseteq \varphi(H) \wedge \varphi(K) \in \Sigma,$$

so that

$$\varphi(H \cap K) \supseteq \varphi(H) \wedge \varphi(K).$$

Conversely if $H \cap K \supseteq W \in \Sigma$, then

$$H \supseteq H \cap K \supseteq W \in \Sigma,$$

so that

$$\varphi(H) \supseteq W,$$

and similarly

$$\varphi(K) \supseteq W.$$

Hence

$$\varphi(H) \wedge \varphi(K) \supseteq W,$$

or

$$\varphi(H) \wedge \varphi(K) = \varphi(H \cap K).$$

Thus the consideration of a meet-homomorphism is equivalent to the study of a subset Σ of $L(G)$ satisfying the above conditions (1) and (2).

We shall consider a special case in which the above mapping defines an L-homomorphism of G.

Theorem 11. (ZAPPA [8].) *Let N be a proper subgroup of a finite group G. Then the mapping φ defined by $\varphi(U) = U \cap N$ is an L-homomorphism if and only if N is a normal subgroup of G and there exist three subgroups M, L and H such that*

(1) *their orders are relatively prime to each other,*

(2) *M and L are normal subgroups of G,*

(3) *$M \cap (L\,H) = e$, $L\,H \supseteq N \supseteq L$,*

(4) *H is nilpotent and each SYLOW subgroup of H is either cyclic or a generalized quaternion group,*

(5) *$H \cap N = K$ is cyclic, and if H is not cyclic, the order of K is twice an odd number, and*

(6) *if a subgroup or an element of L is permutable with all elements of $N \cap S$ where S is a SYLOW subgroup of H, then it is permutable with all elements in S.*

Proof. Suppose that the mapping $\varphi(U) = U \cap N$ is an L-homomorphism. Then N is the u-kernel of φ, and hence a normal subgroup of G. We shall show that there exist three subgroups M, L and H satisfying conditions (1) to (6). To do this we consider a slightly more general situation.

Suppose that φ is an arbitrary L-homomorphism of G onto a lattice L, and N is the u-kernel of φ. Then by (3.6) of p. 70 there exist a normal subgroup M and a subgroup J of G such that M and J have relatively prime orders and

$$G = M\,J,\ M \cap J = e \text{ and } J \supseteq N.$$

Now let Π be the totality of prime numbers p such that φ induces a proper L-homomorphism of a p-SYLOW subgroup S and $\varphi(S)$ is a chain. For every $p \in \Pi$, p-SYLOW subgroups are by (3.4) either cyclic or generalized quaternion groups, and J contains the normal p-SYLOW complement N_p by (3.5) and (3.7) of p. 71. Let L be the intersection of all N_p;

$$L = \cap_{p \in \Pi} N_p.$$

Then $L \subseteq N$ and L is a normal subgroup of G. By a theorem of SCHUR, J contains a subgroup H such that

$$L\,H = J \text{ and } L \cap H = e.$$

H is then isomorphic with J/L, and so nilpotent. It is clear that M, L and H satisfy the conditions (1) to (5).

Suppose now that there are subgroups U and V of L such that $U \supset V$ and $\varphi(U) = \varphi(V)$. Without loss of generality, we may assume that V is the u-kernel of the restriction of φ on U. By this assumption V is a normal subgroup of U. If p is a prime divisor of the order of U/V, the restriction of φ upon a p-SYLOW subgroup T of J is not a projectivity. p is not a prime number in Π, so that by definition of Π, $\varphi(T)$ is not a chain. Hence by (3.4) $p = 2$, and T is a generalized quaternion group. We have assumed that $\varphi(U) = \varphi(V)$ for some $U \supset V$. Hence the index $(U:V)$ is 2 and $U = V \times Z$, where $Z = U \cap T$ is a central subgroup of G with order 2. Note that V is of odd order.

We shall now prove the necessity of the last condition (6). Let S be a SYLOW subgroup of H. Assume first that a subgroup X of L commutes with all elements in $N \cap S$. Then $X(N \cap S) = (N \cap S)\,X$, and so by the modular identity we have

$$\big((N \cap S) \cup X\big) \cap L = X \cup \big((N \cap S) \cap L\big) = X.$$

Since $\varphi(S) = \varphi(N) \cap \varphi(S) = \varphi(N \cap S)$, we have

$$\varphi(X) = \varphi\big(((N \cap S) \cup X) \cap L\big) = \varphi((S \cup X) \cap L).$$

If $(S \cup X) \cap L = U \supset X$, then $U = X \times Z$ where Z is in the center of G. Since both X and S are of odd order, $S \cup X$ can not contain any central subgroup of order 2. Hence

$$(S \cup X) \cap L = X,$$

or X is a normal subgroup of $S \cup X$; in other words X commutes with all elements of S.

Assume now that an element a of L commutes with all elements of $N \cap S$. Then

$$a\,S\,a^{-1} \cap N = a(S \cap N)\,a^{-1} = S \cap N.$$

Applying φ, we conclude that

$$\varphi(S) = \varphi(S \cap N) = \varphi(a S a^{-1} \cap N) = \varphi(a S a^{-1}).$$

Hence

$$\varphi\big((a S a^{-1} \cup S) \cap L\big) = \big(\varphi(a S a^{-1}) \cup \varphi(s)\big) \cap \varphi(L)$$
$$= \varphi(S) \cap \varphi(L) = 0.$$

Hence $a S a^{-1} \cup S$ is nilpotent, and since S is a Sylow subgroup of J, this implies that $a S a^{-1} = S$. Since $\{a\}$ is a normal subgroup of $\{a, S\}$, a commutes with all elements of S. This proves (6).

Suppose now conversely that G and N satisfy the conditions (1) to (6) of this theorem. First of all, take two subgroups U and V of $J = L H$. We shall show that

$$(U \cup V) \cap N = (U \cap N) \cup (V \cap N).$$

J contains a series of normal subgroups

$$N_0 = J \supset N_1 \supset \cdots \supset N_r = N$$

such that each factor group N_{i-1}/N_i is a p_i-group. If H is not cyclic, we take a suitable N_1 so that $p_1 = 2$ and $N_1 \cap H$ is cyclic. Then for each N_i $(i \geq 1)$ and for p_i-Sylow complement L_i of N_i the conditions (5) and (6) are satisfied. Hence we may assume, using induction, that the factor group J/N is of prime power order. Suppose the orders of the p-Sylow subgroups of U, V and N are p^u, p^v and p^n respectively. If $u, v \leq n$ we have $U, V \subseteq N$ and our assertion is trivial. Assume that $u \geq v$ and $u > n$. Take a p-Sylow subgroup U_p of U. $U_p \cap N$ is a p-Sylow subgroup of N. Hence

$$(U \cap N) \cup (V \cap N) = L'(U_p \cap N),$$

where $L' \subseteq L$. Now U_p is contained in a p-Sylow subgroup S_p of J and $S_p \cap N = U_p \cap N$. By (6), every element in S_p is permutable with L'. Hence

$$(L' \cup S_p) \cap N = L'(U_p \cap N).$$

On the other hand, we have $L' \cup S_p \supseteq U$. Take a p-Sylow subgroup V_p of V. Then some conjugate subgroup $a\,(U_p \cap N)\,a^{-1}$ of $U_p \cap N$ contains $V_p \cap N$ where $a \in L'$. Hence

$$a S_p a^{-1} \supseteq V_p \cap N.$$

If $V_p \subseteq N$, we have $L' \cup S_p = L' \cup a S_p a^{-1} \supseteq V$. If $V_p \nsubseteq N$, V_p is contained in a p-Sylow subgroup S_p' of J. Since S_p' and $a S_p a^{-1}$ are p-Sylow subgroups of J, there is an element b in L such that $b\,(a S_p a^{-1})\,b^{-1} = S_p'$. But since

$$a S_p a^{-1} \cap S_p' \supseteq V_p \cap N = S_p' \cap N,$$

b commutes with every element of $S_p' \cap N$. Hence by (6) b commutes with every element of S_p' which implies $a S_p a^{-1} = S_p'$. Hence again we have

$$L' \cup S_p = L' \cup a S_p a^{-1} \supseteq V.$$

In any case we have proved that $L' \cup S_p \supseteq U \cup V$. Hence

$$(U \cup V) \cap N \subseteq (L' \cup S_p) \cap N = L'(U_p \cap N) = (U \cap N) \cup (V \cap N).$$

Combining this with the obvious relation $(U \cup V) \cap N \supseteq (U \cap N) \cup (V \cap N)$, we obtain the required result.

Now the following relation is satisfied for any subgroup U of G:

$$(U \cap N) \cup M = (U \cup M) \cap (N \cup M).$$

Clearly we have $(U \cap N) \cup M \subseteq (U \cup M) \cap (N \cup M)$. On the other hand, indices

$$\big((U \cap N) \cup M : M\big) \text{ and } \big((U \cup M) \cap (N \cup M) : M\big)$$

are equal, so that we have equality.

We shall finally prove that

$$(U \cup V) \cap N = (U \cap N) \cup (V \cap N)$$

for arbitrary subgroups U and V of G. First of all, the above identity implies

$$\big((U \cup V) \cap N\big) \cup M = (U \cup V \cup M) \cap (N \cup M).$$

On the other hand, as $G/M \cong J = H L$, we have

$$\begin{aligned} &(U \cup V \cup M) \cap (N \cup M) \\ &= \big((U \cup M) \cap (N \cup M)\big) \cup \big((V \cup M) \cap (N \cup M)\big) \\ &= \big((U \cap N) \cup M\big) \cup \big((V \cap N) \cup M\big) = \big((U \cap N) \cup (V \cap N)\big) \cup M. \end{aligned}$$

Now since $N \cap M = e$, and since M is normal, we obtain

$$\begin{aligned} (U \cup V) \cap N &\cong \big((U \cup V) \cap N\big) \cup M/M = \big((U \cap N) \cup (V \cap N)\big) \cup M/M \\ &\cong (U \cap N) \cup (V \cap N). \end{aligned}$$

Since $(U \cup V) \cap N \supseteq (U \cap N) \cup (V \cap N)$, we get the required equality.

Corollary. *Let φ be an L-homomorphism of a finite group G. If G_0 is the u-kernel, E the l-kernel of φ and if $E_0 = E \cap G_0$, then the mapping*

$$\theta : U \to (U \cap G_0) \cup E_0$$

defines an L-homomorphism of G onto $L(G_0/E_0)$. Moreover φ is the product of the restriction φ_0 of φ on G_0/E_0 and θ, i. e. $\varphi = \varphi_0\, \theta$ (Zappa [*11*]).

As an application we can determine the *neutral elements* in subgroup lattices (Suzuki [*2*] and Zappa [*10*]). An element l of a lattice L is called neutral if every triple l, x, y of elements in L generates a distributive sublattice. If a group G is L-decombosable, i. e. $G = H \times K$ and $L(G) = L(H) \times L(K)$, then H and K are neutral elements in $L(G)$. Moreover a subgroup U of G is neutral if and only if $U \cap H$ and $U \cap K$ are both neutral in $L(H)$ and $L(K)$ respectively.

Let N be a neutral element of $L(G)$. Then the mappings $U \to U \cup N$ and $U \to U \cap N$ are both homomorphisms. We can apply theorems 6 and 11.

Theorem 12. *Assume that the lattice $L(G)$ of all subgroups of G is irreducible. A proper subgroup N of G is a neutral element in $L(G)$ if and only if N is contained in the center of G, and G contains a normal subgroup M and a subgroup H such that*

(1) $G = M\,H,\ M \cap H = e,\ H \supseteq N,$

(2) *the orders of M and H are relatively prime, and*

(3) *H is nilpotent and its* Sylow *subgroups are either cyclic or a generalized quaternion group.*

The neutral elements in the lattice of normal subgroups have been considered by Greco [*1*] and Zitarosa [*1*]. When a group G is nilpotent, a normal subgroup N of order n in G is neutral in the lattice of normal subgroups if and only if N is the only normal subgroup of order n.

8. Structure of finite groups which admit a proper L-homomorphism.

Theorem 13. *If a finite group G of even order greater than two contains only one element of order 2, then G admits a proper L-homomorphism.*

This theorem follows from theorem 6 of p. 65.

Theorem 14. *Let G be a finite group such that the number of elements of order 2 in G is not equal to 1.*

*G admits a proper L-homomorphism, if and only if there are two normal subgroups N and K of G, r distinct prime divisors $p_1, p_2, \ldots, p_r$ of the order of G and a p_i-*Sylow *subgroup H_i for each $i = 1, 2, \ldots, r$ with the following properties:*

(1) *the orders of N and K are relatively prime,*

(2) $G = N \cup K \cup H_1 \cup H_2 \cup \cdots \cup H_r$,

(3) *$H_1 \cup H_2 \cup \cdots \cup H_r$ is cyclic,*

(4) *each H_i contains a series of subgroups*

$$e = H_{i0} \subset H_{i1} \subset \cdots \subset H_{im_i} \subseteq H_i$$

such that

(4a) *the subgroup* $K \cup H_{1m_1} \cup H_{2m_2} \cup \cdots \cup H_{rm_r}$ *is a normal subgroup of G, and*

(4b) *if U is a subgroup of* H_i *which contains* H_{ij}, *but not* $H_{i,j+1}$, *then a subgroup or an element of K is permutable with every element of U, whenever it is permutable with all elements of* H_{ij}, *and*

(5) *if* $N = e$, *and* $H_{im_i} = H_i$ *for all* $i = 1, 2, \ldots, r$, *then at least one index* $(H_{jk} : H_{jk-1})$ *is not a prime number.*

Proof. Assume first that G admits a proper L-homomorphism φ. By (3.6) of p. 70, G contains a normal subgroup N and a subgroup J such that their orders are relatively prime and $NJ = G$. Let $\Pi = \{p_1, p_2, \ldots, p_r\}$ be the totality of prime divisors of G such that φ induces a proper L-homomorphism of a p_i-Sylow subgroup of G. If $p \in \Pi$, a p-Sylow subgroup of G is by (3.4) either cyclic or a generalized quaternion group. But if it were not cyclic, then G would contain only one element of order 2. This is not the case, so that each p-Sylow subgroup of G is cyclic, if $p \in \Pi$. Hence by (3.5) J contains a p-Sylow complement N_p, which is a normal subgroup of J. Put $K = \bigcap_{p \in \Pi} N_p$. Then by a theorem of Schur, J contains a subgroup H such that $J = HK$ and $H \cap K = e$. H is then isomorphic with J/K and so cyclic. Let H_i be the p_i-Sylow subgroup of H. $\varphi(H_i)$ is a chain, so that we may assume that the dimension of $\varphi(H_i)$ is m_i. Let H_{ij} be the smallest subgroup of H_i such that $\varphi(H_{ij})$ is the j-dimensional element of $\varphi(H_i)$. Then it is clear that the subgroup $K \cup H_{1m_1} \cup \cdots \cup H_{rm_r}$ is the u-kernel of φ and hence normal in G. The necessity of the last condition (4b) may be proved in a similar fashion as in the proof of (6) of theorem 11 (cf. p. 76).

Assume conversely that G satisfies all requirements. Put

$$G_0 = K \cup H_{1m_1} \cup \cdots \cup H_{rm_r}.$$

Then G_0 is a normal subgroup of G by (4a). By theorem 11, the mapping $V \to V \cap G_0$ of $L(G)$ onto $L(G_0)$ is an L-homomorphism (which may be an isomorphism). If Y is a subgroup of G_0, then the order of Y is of the form $k \prod p_i^{e_i}$, where $(k, p_i) = 1$ for all $i = 1, 2, \ldots, r$. We call this $p_i^{e_i}$ the *i-th factor* of Y.

Now let Σ be the totality of subgroups of G_0 whose i-th factors are equal to the order of some subgroup H_{ij} for all $i = 1, 2, \ldots, r$. Then it is easy to see that

(1) $\Sigma \ni e$,

(2) if $X, Y \in \Sigma$, then $X \cup Y \in \Sigma$.

Define $\varphi(V)$ as the maximal subgroup of V, which is a member of Σ. Then we have

$$\varphi(U \cap V) = \varphi(U) \cap \varphi(V).$$

We want to show that φ defines an L-homomorphism of G onto Σ. Since the mapping $V \to V \cap G_0$ is an L-homomorphism, it suffices to prove the equality

$$\varphi(X \cup Y) = \varphi(X) \cup \varphi(Y)$$

for every pair of subgroups of G_0.

Define Σ_i $(i = 0, 1, 2, \ldots, r)$ as the totality of subgroups of G_0, whose j-th factors are equal to the order oi some H_{jk} for all $j \leq i$, and let $\varphi_i(X)$ be the maximal subgroup of X such that $\varphi_i(X) \in \Sigma_i$.

Now G_0 contains a series of normal subgroups

$$K = K_r \subset K_{r-1} \subset \cdots \subset K_1 \subset K_0 = G_0$$

such that each factor group K_{i-1}/K_i is a p_i-group. We shall prove by using induction on i that

$$(X \cup Y) \cap K_i = \big(\varphi_i(X) \cup \varphi_i(Y)\big) \cap K_i$$

holds for every pair of subgroups X and Y of G_0.

If $i = 0$, this is trivial. Put $\varphi_{i-1}(X) = X_0$, $\varphi_{i-1}(Y) = Y_0$ and assume

$$(X \cup Y) \cap K_{i-1} = (X_0 \cup Y_0) \cap K_{i-1}.$$

Then $\varphi_i(X) = \varphi_i(X_0)$, $\varphi_i(Y) = \varphi_i(Y_0)$ and if $T = (\varphi_i(X) \cup \varphi_i(Y)) \cap K_i$, then $(X_0 \cup Y_0) \cap K_i \supseteq T$.

Take a p_i-Sylow subgroup X_p of X_0, and a p_i-Sylow subgroup Y_p of Y_0. We may assume, without loss of generality, that the order of X_p is greater than that of Y_p. $X'_p = \varphi_i(X_p) = \varphi(X_p)$ is a p_i-Sylow subgroup of $\varphi_i(X_0) = \varphi_i(X)$ by definition. Since the order of X_p is greater than the order of Y_p, X'_p is also a p_i-Sylow subgroup of $\varphi_i(X) \cup \varphi_i(Y)$. Hence $\varphi_i(X) \cup \varphi_i(Y)$ contains a subgroup Z' such that

$$Z'\, T = \varphi_i(X) \cup \varphi_i(Y),\ Z' \cap K_i = e \text{ and } Z' \supseteq X'_p.$$

Similarly $X_0 \cup Y_0$ contains a subgroup Z such that

$$X_0 \cup Y_0 = ((X_0 \cup Y_0) \cap K_i)\, Z,\ Z \cap K_i = e \text{ and } Z \supseteq X_p.$$

Since G_0/K is solvable, there exists an element a of K such that $Z_0 = a\, Z\, a^{-1} \supseteq Z'$. The element a is therefore permutable with every element of X'_p, so that by (4b) a commutes with all elements of X_p. Hence $Z_0 \supseteq X_p \cup Z'$ or $Z_0 = X_p \cup Z'$. Now $Z'\, T = T\, Z'$, and again using (4b) we conclude that Z_0 commutes with T. Hence

$$(T \cup Z_0) \cap K_i = T \cup (Z_0 \cap K_i) = T.$$

If k is an integer less than i, then every p_k-Sylow subgroup of X_0 or Y_0 is transformed into Z_0 by a suitable element of T. Hence in particular X_0 is contained in $T\, Z_0$. Take any p_i-Sylow subgroup Y_p of Y_0.

$Y'_p = \varphi_i(Y_p)$ is a p_i-subgroup of $\varphi_i(Y_0)$, and hence there exists an element x of T such that $x X_p x^{-1} \supseteq x X'_p x^{-1} \supseteq Y'_p$. Now Y_p is contained in a p_i-SYLOW subgroup X''_p of $X_0 \cup Y_0$. Hence there is an element y of K such that $y(x X_p x^{-1}) y^{-1} = X''_p$. y commutes with all elements of Y'_p, and so by (4b) y commutes with all elements of Y_p. This implies that $x X_p x^{-1} \supseteq Y_p$, or $Y_p \subseteq T Z_0$. Hence we have

$$T Z_0 \supseteq X_0 \cup Y_0.$$

Hence

$$T = (T \cup Z_0) \cap K_i \supseteq (X_0 \cup Y_0) \cap K_i \supseteq T,$$

and so

$$\begin{aligned}(X \cup Y) \cap K_i &= (X \cup Y) \cap K_{i-1} \cap K_i = (X_0 \cup Y_0) \cap K_{i-1} \cap K_i \\ &= (X_0 \cup Y_0) \cap K_i \\ &= \big(\varphi_i(X) \cup \varphi_i(Y)\big) \cap K_i.\end{aligned}$$

If we let $i = r$, we get

$$(X \cup Y) \cap K = \big(\varphi(X \cup Y)\big) \cap K = \big(\varphi(X) \cup \varphi(Y)\big) \cap K.$$

Since the i-th factors of $\varphi(X \cup Y)$ and $\varphi(X) \cup \varphi(Y)$ are equal, we conclude that

$$\varphi(X \cup Y) = \varphi(X) \cup \varphi(Y).$$

Hence the mapping φ defines an L-homomorphism of G onto Σ, and φ is proper by (5).

9. L-homomorphisms onto a nilpotent group.

If we consider an L-homomorphism of G onto a subgroup lattice of another specified group G', we can obtain more precise results. Several special cases have been studied by PERMUTTI [*1*], ZACHER [*1, 2, 3*], ZAPPA [*6, 11*] and SUZUKI [*2*].

Let φ be an L-homomorphism of a finite group G onto the subgroup lattice $L(G')$ of another group G'. Denote by E and G_0 the l- and u-kernel of φ. They are normal subgroups. By (3.6) G contains a normal subgroup N and a subgroup H such that

$$G = N H,\ N \cap H = e,\ H \supseteq G_0 \text{ and } E = E_0 \times N$$

where $E_0 = E \cap G_0$. If φ induces a proper L-homomorphism of G_0/E_0 onto $L(G')$, G_0 contains by (3.8) a normal subgroup G_1 such that φ induces a projectivity of G_1/E_0 into $L(G')$ and G_0/G_1 is cyclic. These notations we shall use throughout this section.

Proposition 3.9. *Suppose G' to be a p-group. If G' is neither cyclic nor a P-group, H is also a p-group and coincides with G_0. G is therefore a* *direct product of N and G_0. If G' is a P-group, H is either a p-group or an upper semi-modular group of order $p^m q^n$, where q is a prime number less than p, and G_0 is its maximal normal M-group.*

Proof. We shall assume that G' is not cyclic. If H/G_0 is not cyclic, H is a direct product of its 2-SYLOW subgroup S_2 and 2-SYLOW complement H_2. Hence H is L-decomposable, so that $L(G')$ is also decomposable which is impossible, since G' is a p-group. Hence H/G_0 is cyclic.

Now φ induces a projectivity of G_1/E_0 into $L(G')$. Since the image group is a p-group, G_1/E_0 is either a P-group or of prime power order (I. theorem 12, p. 12). If G_1/E_0 were a non-abelian P-group, φ would induce a projectivity of V/E_0 into G', where V is a subgroup of G_0 which covers G_1. This is a contradiction since the order of V/E_0 would be divisible by three distinct primes (see (3.8), p. 73, and I. theorem 12, p. 12). Hence in any case H is a p-group or a group of order $p^m q^n$ $(p > q)$. If H is a p-group, it follows from (3.4), p. 70, that H coincides with G_0.

Assume next that H is of order $p^m q^n$. G_0/E_0 is a group of order $p^\alpha q^\beta$ and its p-SYLOW subgroup $\overline{S}$ is normal. Now φ induces a projectivity of $\overline{S}$ into G'. If we take a subgroup $\overline{T}$ of G_0/E_0, which covers $\overline{S}$, φ also induces a projectivity of $\overline{T}$. Hence $\overline{T}$ is a P-group. Let $\overline{Q}$ be one of the q-SYLOW subgroups of G_0/E_0. Consider a subgroup $\overline{V}$ covering $\overline{Q}$. Since G' is a p-group, $\varphi(\overline{S}) \cap \varphi(\overline{V})$ is of prime order. Hence $\overline{S} \cap \overline{V}$ is a normal subgroup of order p. It is now easy to see that $\varphi(\overline{V})$ is a P-group of dimension 2. This implies that

$$\varphi(\overline{T}) = \varphi(\overline{S} \cup \overline{V}) = \varphi(G_0/E_0), \text{ or } G_0/E_0 = \overline{T}.$$

Hence G_0/E_0 and G' are both P-groups.

Since SYLOW p-complements of H are not normal, the orders of H/G_0 and E_0 are both powers of q. The p-SYLOW subgroup S of H is clearly normal in H and φ induces a projectivity of S into G'. Take any subgroup V of order p and any q-SYLOW subgroup Q of H. Then $\varphi(V \cup Q)$ is a P-group of order p^2. Hence $(V \cup Q) \cap S$ is of prime order and hence coincides with V;

$$(V \cup Q) \cap S = V.$$

This implies that V is a normal subgroup of H. Put $Q = \{b\}$; then for any element a of S we have

$$b\,a\,b^{-1} = a^x,\ x \not\equiv 1,\ x^{q^t} \equiv 1 \pmod{p}.$$

Hence H is an UM-group and G_0 is its maximal normal M-group.

On the basis of theorems 3 (p. 60), 11 (p. 75), (3.6) (p. 70) and (3.9) we can prove the following theorem (SUZUKI [2], p. 382, theorem 8).

Theorem 15. *There exists an L-homomorphism of the group G onto a nilpotent group* $G' = \prod_{i=1}^{t} S_i$, *where* S_i *is a* p_i-Sylow *subgroup of* G', *if and only if there exist a normal subgroup N and a subgroup H of G with the following properties:*

(1) $NH = G$ *and* $N \cap H = e$,

(2) *the order of N is relatively prime to that of H,*

(3) *H is a direct product of groups* H_i $(i = 1, 2, \ldots, t)$ *having mutually prime orders:* $H = \prod_{i=1}^{t} H_i$,

(4) *if* S_i *is a cyclic group of order* $p_i^{e_i}$, H_i *is a cyclic group of prime power order or a generalized quaternion group, and* H_i *contains a normal subgroup* K_i *of G such that* K_i *is cyclic, its order is* $q_i^{f_i}$ *and* $f_i \geq e_i$, *and if* H_i *is not cyclic we have* $e_i = f_i = 1$,

(5) *if* S_k *is a P-group of order* p_k^{n+1} $(n \geq 1)$, H_k *is either isomorphic to* S_k, *or a quaternion group* $(n = 1, p_k = 2)$, *or an UM-group of order* $p_k^n q^m$ (*q is a prime and* $p_k > q$), *and its maximal normal M-group is a normal subgroup of G, and*

(6) *if* S_l *is neither cyclic nor a P-group,* H_l *is also a p-group and normal in G. In this case if* $L(H_l)$ *is not isomorphic to* $L(S_l)$, H_l *is a generalized quaternion group and* S_l *is isomorphic to the factor group* H_l/Z_l *modulo its center* Z_l.

Returning to the general case we prove

Proposition 3.10. $\varphi(G_1)$ *is a normal subgroup of* G'.

Proof. In changing the notations, we shall assume that the *u*-kernel of φ coincides with G and that the *l*-kernel of φ coincides with e. Take a p-Sylow subgroup S of G in which φ induces a proper L-homomorphism. By (3.4) and (3.5), p. 70, S must be cyclic, and G has a normal Sylow p-complement N. We shall first prove that $\varphi(S)$ is also a Sylow subgroup of G.

Since $\varphi(S)$ is a cyclic group of prime power order, it is contained in some Sylow subgroup S' of G'. Take the greatest subgroup U of G such that $\varphi(U) = S'$. Then U clearly contains S. If S' were a P-group, $\varphi(S)$ would be of prime order. On the other hand, taking the maximal subgroup M of S, we have $\varphi(M) \neq \varphi(S)$, as the *u*-kernel of φ coincides with G. Hence we could have $\varphi(M) = e$, that is, M would be contained in the *l*-kernel of φ and by our assumption $M = e$. Hence S is mapped isomorphically onto $\varphi(S)$, contrary to our assumption. Hence S' is not a P-group and U is also of prime power order by (3.9). Hence U must coincide with S, that is, $S' = \varphi(S)$.

Next we shall prove that $S' = \varphi(S)$ is contained in the center of its normalizer. Take a subgroup V' of G' such that S' is normal in V' and V'/S' is of prime power order, say of order q^n (q is a prime number).

Take a subgroup V of G such that $\varphi(V) = V'$; then $\varphi(V \cap N)$ is a q-SYLOW subgroup Q' of V'. If $V \cap N$ is cyclic, but not of the same order as Q', S is normal in V by (3.5), and hence V and also V' are directly decomposable.

We may assume that φ induces a projectivity of $V \cap N$. Since the l-kernel of φ coincides with e, φ induces a projectivity of a subgroup T of V, covering $N \cap V$. Let $\varphi(T) = T'$, then by our assumption $T' \cap S'$ is normal in T'. If $T \cap S$ were not normal in T, T would be a P-group which would imply that Q' has prime order. Hence $V \cap N$ would also be of prime order. Since $\varphi(S)$ is normal in V', V' is a P-group, which leads us to the same contradiction as above. Hence $T \cap S$ is a normal subgroup of T, and so T is a direct product of $N \cap V$ and $T \cap S$. This implies that $T \cap S$ is normal in V. If S were not normal in V, there would be another p-SYLOW subgroup S^* of V. S^* would also contain $T \cap S$. Hence we would have $\varphi(S^*) \cap S' \neq e$. Since $\varphi(S^*)$ is a cyclic group of prime power order, this gives a contradiction. Hence we have

$$V = (N \cap V) \times S \text{ and } V' = Q' \times S'.$$

S' is thus contained in the center of its normalizer and G' contains a normal subgroup N' such that $N' S' = G$ and $N' \cap S' = e$ by a theorem of BURNSIDE.

We shall now prove that $\varphi(N) = N'$. Take all p-SYLOW subgroups $S = S_1, S_2, \ldots, S_t$ of G. Then φ induces a proper L-homomorphism in each S_i. Hence the $\varphi(S_i)$ are SYLOW subgroups of G' and are contained in centers of their normalizers, as proved above. G' contains therefore SYLOW complements $N' = N_1', N_2', \ldots, N_r'$ which are normal in G'. Put $D' = \bigcap_{i=1}^{r} N_i$. Take a subgroup D of G such that $\varphi(D) = D'$. Since $D' \cap \varphi(S_i) = e$ $(i = 1, 2, \ldots, t)$, we have $D \cap S_i = e$ $(i = 1, 2, \ldots, t)$, which implies that the order of D is prime to p, or $D \subseteq N$. Since $\varphi(N) \supseteq D'$, $\varphi(N) \cap \varphi(S) = e$ and $\varphi(N) \cup \varphi(S) = G'$, we have $\varphi(N) = N'$. This proves our proposition.

This proposition implies in particular the following theorem.

Theorem 16. *Let G be a finite solvable group, and φ be an L-homomorphism of G onto another group G'. Then G' is also solvable.*

Chapter IV.

Dualisms of subgroup lattices.

1. Dualisms (of abelian groups).

A dual-isomorphism between subgroup lattices is called a *dualism* of subgroup lattices. Let φ be a dualism of the group G onto H. Then the lattice $L(G)$ of subgroups of G is mapped onto $L(H)$ and for every pair of subgroups U and V of G we have

$$\varphi(U \cap V) = \varphi(U) \cup \varphi(V),$$

and

$$\varphi(U \cup V) = \varphi(U) \cap \varphi(V).$$

H is called a *dual of* G.

Not every group has a dual, for instance the quaternion group has none.

Proposition 4.1. (BAER [5].) *If a group G has a dual, then G is a torsion group.*

Proof. Let a group H be a dual of G and φ be a dualism of G onto H. By way of contradiction suppose that G contains an infinite cyclic subgroup $Z = \{u\}$. If Z^n is the cyclic subgroup generated by u^n, then

$$Z = Z^1 \supset Z^2 \supset \cdots \supset Z^{2^i} \supset \cdots$$

is a descending chain of subgroups and $\bigcap_i Z^{2^i} = e$. Put $W_i = \varphi(Z^{2^i})$. Then $W_0 \subset W_1 \subset \cdots$ is an ascending chain such that $\bigcup_i W_i = H$. Consider the subgroup $V = \varphi(Z^3)$. Since Z covers Z^3, W_0 is a maximal subgroup of V. Now there must be an integer k such that $W_k \supseteq V$ since $\bigcup_i W_i = H$. Taking the inverse map φ^{-1}, we have

$$\varphi^{-1}(W_k) = Z^{2^k} \subseteq Z^3 = \varphi^{-1}(V),$$

which is a contradiction. This proves our proposition.

In case of finite abelian groups, we can assert the existence of duals.

Proposition 4.2. (BAER [3].) *A finite abelian group G is self-dual,* i.e. *there exists an auto-dualism φ such that*

$$U \cong G/\varphi(U) \text{ and } \varphi(U) \cong G/U \text{ for any } U \subseteq G.$$

Proof. Consider the totality G^* of homomorphisms of G into the real numbers modulo 1. G^* forms an abelian group, isomorphic with G, under the usual multiplication. There exists an isomorphism θ of G^* onto G, which induces projectivity of $L(G^*)$ upon $L(G)$. For any subgroup U of G, the totality of homomorphisms such that $\sigma(u) = 1$ $(u \in U)$ forms a subgroup U^* of G^* and it is easy to see that $U^* \cong G/U^*$. Then $\varphi(U) = \theta(U^*)$ is the desired auto-dualism φ.

In case of infinite groups we have however

Proposition 4.3. (BAER [5].) *An infinite abelian group G of exponent a prime p has no dual.*

Proof. Suppose on the contrary that G has a dual H and denote by φ the dualism of G onto H. Take any element $u \neq 1$ of H, then the factor group of G by the inverse image of $U = \{u\}$ is a finite cyclic group, and so of order p. This implies that u is of prime order. Take another element v of H such that $\{u\} \cap \{v\} = e$. Then

$$W = \varphi^{-1}\{u, v\} = \varphi^{-1}\{u\} \cap \varphi^{-1}\{v\}$$

and G/W is an abelian group of type (p, p). Hence $\{u, v\}$ is a P-group which implies that all elements of order p in H form an abelian subgroup P of H with finite index. Now φ induces a dualism of $P' = G/\varphi^{-1}(P)$ onto P. Since $\varphi^{-1}(P)$ is a finite group, P' is not of finite order by hypothesis. Suppose P' has $\aleph$ subgroups of order p. Then the (cardinal) number of subgroups in P' with index p is equal to $2^{\aleph}$, which is the number of subgroups of order p in P. Hence P has $2^{2^{\aleph}}$ subgroups of index p. By a dual mapping P' has $2^{2^{\aleph}}$ subgroups of order p. Hence we have $\aleph = 2^{2^{\aleph}}$ which is impossible.

Theorem 1. *An abelian group G has a dual if and only if G is a torsion group and every primary component of G is a finite group* (BAER [5]).

Proof. Suppose G has a dual. Then G is a torsion group by (4.1). Hence G is the direct product of its primary components; $G = \prod G_p$. If G were to contain a subgroup Z of type p^∞, Z would be a direct factor of G. Hence $G = Z \times Z'$, and $G/Z' \cong Z$ would have a dual, which is impossible. Hence G does not contain a subgroup of type p^∞. Now $G_p/G_p{}^p \cong G/G^p$ has a dual, so that by (4.3), $G_p/G_p{}^p$ is of finite order. Hence G_p itself is a finite group.

The converse follows from (4.2) and I. theorem 4.

2. Nilpotent groups with duals.

The structure of nilpotent groups with duals has been determined completely (SUZUKI [*1*]). Here we call a group G nilpotent, if G has a central series of finite length joining G and e. We need the following lemma (ZASSENHAUS [*1*], p. 115).

Lemma 1. *A nilpotent group is finite if and only if its commutator-factor group is finite.*

Theorem 2. *A nilpotent group G has a dual if and only if*

(1) *G is a torsion group,*

(2) *every primary component of G is a finite M-group which is not a Hamiltonian 2-group.*

Proof. We shall assume that G has a dual H and denote by φ the dualism of G onto H. The first condition (1) follows from (4.1). Hence, as G is nilpotent, G is a direct product of its primary components G_p, that is, $G = \prod G_p$. We have then $L(G) = \prod L(G_p)$ by I. theorem 4.

G has a dual if and only if every G_p has a dual. Hence we may assume G itself to be primary. Denote the commutator subgroup of G by $C(G)$ and put $\varphi(C(G)) = K$. $G/C(G)$ and K are duals to each other by φ. Since $G/C(G)$ is a primary abelian group, $G/C(G)$ must be finite by theorem 1. Hence, by the above lemma, G itself is finite.

We shall prove next that if a finite p-group G has a dual, G is an M-group. Take the Φ-subgroup Φ of G. Then $\varphi(\Phi) = N$ is clearly normal in H and is a P-group. If N is a p-group, H is also a p-group. $L(H)$ is therefore lower semi-modular, and at the same time upper semi-modular as an image of $L(G)$ by φ. $L(H)$ is then a modular lattice by a theorem of BIRKHOFF (Cf. BIRKHOFF [1], p. 43). Hence G is an M-group.

If N is not a p-group, its order is $p^n q$, where p and q are primes ($p > q$). Since two subgroups of order q in N do not generate a q-group, a q-SYLOW subgroup Q of H is either a cyclic group or a generalized quaternion group. As H is a J-group, the p-SYLOW subgroup S of H is normal and covers a normal subgroup T of H by I. theorem 9. Take two subgroups U and V of G such that $\varphi(U) = S$ and $\varphi(V) = T$. It is clear that V and H/T are duals of each other and that U and H/S are also duals of each other. Hence Q, isomorphic to H/S, is cyclic. U is then cyclic and V is an M-group. (Note that $p > q \geq 2$.) Hence $L(V)$ is clearly self-dual, and the subgroup lattices of V and H/T are isomorphic. By I. theorem 12, H/T must be a P-group. This implies that Q is contained in N. Since $L(H)$ has no reducible interval, we have $H = N$. Hence H is again an M-group. We obtain thus the first part of our theorem.

If G is a finite modular p-group which is not a Hamiltonian 2-group, there exists a projectivity of G upon an abelian group by II. theorem 7 (p. 39). Hence $L(G)$ is clearly self-dual. This completes our proof.

The generalized nilpotent groups may be defined in various ways. Here we define a (generalized) nilpotent group by the property that $\cap_i N_i = e$, where N_i are the higher commutator subgroups defined by $N_0 = G$, and inductively $N_i = [G, N_{i-1}]$. As in the usual case, a nilpotent group is a direct product of its primary components, if it is a torsion group.

The above theorem 2 is still valid if we consider generalized nilpotent groups in the sense defined above. Suppose that a generalized nilpotent group G admits a dualism φ. As before we may assume that G is primary. We want to prove that G is finite. Let N_i be the i-th higher commutator subgroup of G. Then G/N_i is a nilpotent group in the usual sense. Since G/N_i has a dual, it is a finite M-group by theorem 2. Since every finite M-group is metabelian, the commutator subgroup $A = N_1$ is abelian. We take one element from each coset of G modulo A. Let $\Sigma = \{a, b, \ldots\}$ be the totality of these representatives. Since G/N_1 is a finite group, Σ is a finite set. By definition we have clearly $G = \{\Sigma, A\}$. Let U be the subgroup generated by Σ. If $U \neq G$, $\varphi(U)$ is a proper subgroup of $\varphi(G)$. Since $\varphi(G)$ is a torsion group by (4.1), $\varphi(U)$ contains a minimal subgroup V. Then $\varphi^{-1}(V) = M$ is a maximal subgroup of G, such that $M \supseteq U$. Set $D = M \cap A$. D is a normal subgroup of both M and A, and hence of G. Consider the factor group G/D. $\bar{A} = A/D$ is then a minimal normal subgroup of G/D, $\bar{A}$ is generated by each of its elements $a \neq 1$ and its conjugates. Since G/A is finite, $\bar{A}$ is generated by a finite set of elements. Since $\bar{A}$ is a torsion group and abelian, $\bar{A}$ is a finite group. Hence G/D itself is finite. Since G/D is a finite p-group, M is a normal subgroup of G. Hence we have $M \supseteq A$ contradicting the definition of M. Hence U must coincide with G. We have proved that G has a finite set of generators, and since A has finite index, A is also generated by a finite set of elements. A is, then, a finite group and this completes the proof.

The following two theorems are immediate consequences of the above considerations.

Theorem 3. *A nilpotent group G and a group H are duals of each other if and only if $L(G)$ is self-dual and $L(H) \cong L(G)$.*

Theorem 4. *If a nilpotent group G has a dual, then there exists a projectivity of G upon an abelian group.*

3. Finite solvable groups with duals.

In this section we consider only finite solvable groups.

Theorem 5. *A finite solvable group G has a dual if and only if it is a direct product of groups G_i of mutually prime orders, each direct factor G_i being either a P-group or an M-group of prime power order which is not a Hamiltonian 2-group.*

The proof of this theorem may be divided into two parts. First we shall prove the following proposition.

Proposition 4.4. *If a solvable group has a dual, it is super-solvable.*

Proof. Let G be a solvable group with a dual H, and let φ be the dualism of G onto H. We shall prove this lemma by induction on the order of G.

As G is solvable, G contains a normal subgroup N with prime index, say p. $N^* = \varphi(N)$ is then a minimal subgroup of H. Suppose that N^* is normal. Then N and H/N^* are duals to each other and hence by the inductive hypothesis H/N^* is a J-group, so that H and, at the same time, G is a J-group. (Cf. I. theorem 9, p. 9.)

If N^* is not normal in H, there exists an auto-projectivity ψ of G such that $\psi(N) = N_1 \neq N$. If N_1 is not normal, G is an S-group by

(2.14). Hence

$$G = P \times G_1 \text{ and } L(G) = L(P) \times L(G_1),$$

where P is a P-group. G_1 has clearly a dual and is therefore, by the inductive hypothesis, a J-group. Since a P-group is also a J-group, G is a J-group.

Now we shall assume that N_1 is normal. If the index $(G:N_1) = q$ is not equal to p, a p-SYLOW subgroup of G is mapped by ψ onto a q-SYLOW subgroup. Hence by II. theorem 11, G contains a normal subgroup D such that D is invariant under all auto-projectivities of G and the factor group G/D is cyclic. $D^* = \varphi(D)$ is then a normal subgroup of H and is cyclic. We may apply the inductive hypothesis on D and H/D^*, to prove our assertion.

It remains to prove our lemma in the case $(G:N_1) = p$. Put

$$K = \cap \psi(N),$$

where ψ runs through all auto-projectivities of G. We may assume that each $\psi(N)$ is a normal subgroup of G with index p. Under this assumption G/K is an abelian P-group. $K^* = \varphi(K)$ is clearly normal in H. Since K and H/K^* are duals to each other, K is a J-group by the hypothesis of induction. Let r be the largest prime factor of the order of K. Then an r-SYLOW subgroup N_0 of K is a normal subgroup of G. If N_0 is cyclic, we can prove (4.4) in a way similar to the above. Suppose that N_0 is not cyclic. If $N_0^* = \varphi(N_0)$ is not normal in H, G is an S-group and our proposition follows by induction.

Now we may assume that N_0^* is normal in H. Then H/N_0^* is a p-group or a P-group by I. theorem 12. If $r \neq p$, the extension over N_0 by G/N_0 splits; that is, there exists a subgroup Q of G such that $Q\,N_0 = G$ and $Q \cap N_0 = e$. Since H/N_0^* is a p-group or a P-group, there exists a normal subgroup U of N_0 such that $\varphi(U)$ is a normal subgroup of H with prime index. Then $\varphi(Q)$ covers $\varphi(Q) \cap \varphi(U)$. Hence $U \cup Q$ covers Q. We have therefore $U = (U \cup Q) \cap N_0$ which implies that U is a normal subgroup of prime order. Our proposition follows then immediately. If $r = p$, a p-SYLOW subgroup P of G is normal by the hypothesis of induction. We can prove (4.4) in a way similar to the above, replacing N_0 by P.

Proposition 4.5. *If a solvable group has a dual, it is an M-group.*

Proof. Let G be a solvable group with a dual H, and let φ be the dualism of G onto H. We shall again use induction on the order of G. Let p be the largest prime factor of the order of G and P its p-SYLOW subgroup. P is normal by (4.4). If $P^* = \varphi(P)$ is not normal, G is an S-group by theorem 14 (d) on p. 50; that is, $G = T \times G_1$, where T is a P-

group. As G_1 has a dual, it is an M-group by the inductive hypothesis, which implies that G is an M-group.

Now we may assume that P^* is normal in H. By the inductive hypothesis $\overline{G} = G/P$ is an M-group, which is a direct product of groups $\overline{U}_i$ $(i = 1, 2, \ldots, s)$, having mutually prime orders. $\overline{U}_i$ is either a P-group or of prime power order. Take a subgroup U_i $(i = 1, 2, \ldots, s)$ of G, corresponding to $\overline{U}_i$ by the natural homomorphism from G onto $\overline{G}$.

First we shall assume that $s = 1$. In order to prove $\Phi(G) \supseteq \Phi(P)$, take any maximal subgroup M of G. $T = M \cap P$ is clearly a normal subgroup of G, which implies that $T\,\Phi(P) = P$ or T. The former equality implies that $T = P$ and hence we have $T \supseteq \Phi(P)$ or $\Phi(G) \supseteq \Phi(P)$.

Assume now $\Phi(P) = e$. There is a subgroup V of G such that $V\,P = G$ and $V \cap P = e$. Take any minimal subgroup U of P. Then we have

$$\varphi(U) = \big(\varphi(U) \cap \varphi(V)\big) \cup P^* \text{ and } U = (U \cup V) \cap P,$$

which implies that U is normal in G. Applying the hypothesis of induction to the groups G/U and $\varphi(U)$, we see that G/U is a P-group or a direct product of P and V. This proves (4.5) in case $\Phi(P) = e$.

If next $\Phi(P) \neq e$, $G/\Phi(P)$ is a P-group or L-decomposable as proved above. If it is L-decomposable, $G/\Phi(G)$ is clearly also L-decomposable. Hence by (1.1) G is L-decomposable, which proves our proposition. Suppose $G/\Phi(P)$ is a P-group. If P is cyclic, our lemma follows immediately. If P were neither cyclic nor a P-group, H/P^* would be a P-group. By our assumption $\varphi(\Phi(P))$ is a P-group, and hence P^* is also of order p, since P^* is normal. Hence H would be a p-group and G would be a P-group contrary to our assumption. (4.5) is thus proved in case $s = 1$.

If $s \geq 2$, $U_i^* = \varphi(U_i)$ are all normal in H. Since U_i and H/U_i^* are duals to each other, U_i is an S-group. If two groups U_i and U_j were both P-groups, P^*/U_i^* would have the same order as P^*/U_j^*, which is clearly a contradiction. Hence at most one of the U_i is a P-group. This completes our proof of (4.5).

Theorem 5 is now obvious. As a corollary to theorem 5, we have

Corollary. *If a finite solvable group G has a dual, there exists a projectivity of G upon an abelian group.*

It is not yet known whether the assumptions of the finiteness or the solvability are necessary for the validity of these propositions.

We shall denote by $Z(K)$ the centralizer of a subset K of a group G. If U and V are subgroups of G, then

$$Z(U) \supseteq Z(V) \text{ if } U \subseteq V,$$

and

$$Z(Z(U)) \supseteq U.$$

If we assume that every subgroup is a centralizer, then $Z(Z(U)) = U$ and the mapping $U \to Z(U)$ defines a dualism of G upon itself. GASCHÜTZ [1] obtained the following result using theorem 5.

Theorem 6. *Every subgroup of a finite group G is a centralizer, if and only if G is a direct product of P-groups G_i of mutually prime orders of the form $p_i q_i$ (p_i and q_i are different primes).*

Bibliography.

BAER, R.: [1] Situation der Untergruppen und Struktur der Gruppe. S. B. Heidelberger Akad. Wiss. Abhandlungen **2** (1933) 12—17. — [2] Der Kern, eine charakteristische Untergruppe. Compositio Math. **1** (1934) 254—283. — [3] Dualism in abelian groups. Bull. Amer. Math. Soc. **43** (1937) 121—124. — [4] The applicability of lattice theory to group theory. Bull. Amer. Math. Soc. **44** (1938) 817—820. — [5] Duality and commutativity of groups. Duke Math. Journal **5** (1939) 824—838. — [6] The significance of the system of subgroups for the structure of the group. Amer. Journal of Math. **61** (1939) 1—44. — [7] A unified theory of projective spaces and finite abelian groups. Trans. Amer. Math. Soc. **52** (1942) 283—343. — [8] A theory of crossed characters. Trans. Amer. Math. Soc. **54** (1943) 103—170. — [9] Automorphism rings of primary abelian operator groups. Ann. of Math. (2) **44** (1943) 192—227. — [10] Crossed isomorphisms. Amer. Journal of Math. **66** (1944) 341—404.

BAEVA, N. V.: [1] Completely factorizable groups. Doklady Akad. SSSR. **92** (1953) 877—880.

BEAUMONT, R. A.: [1] Projections of the prime power abelian group of order p^m and type $(m - 1, 1)$. Bull. Amer. Math. Soc. **48** (1942) 866—870. — [2] Projections of non-abelian groups upon abelian groups containing elements of infinite order. Amer. Journal of Math. **64** (1942) 115—136.

BERLINKOV, M. L.: [1] Groups possessing a compact lattice of subgroups. Doklady Akad. SSSR. **82** (1952) 505—508. — [2] Groups having a compact lattice of subgroups. Mat. Sbornik **34** (1954) 473—498.

BIRKHOFF, G.: [1] Lattice theory. Coll. Publication Amer. Math. Soc. (1941). Rev. ed. (1948).

BURNSIDE, W.: [1] Theory of groups of finite order. (1911) 2nd ed.

DIAO, A. F.: [1] A theorem about the lattice of subgroups of a group. Gaz. Mat., Lisboa **9** (1948) 18—19.

GASCHÜTZ, W.: [1] Gruppen deren sämtliche Untergruppen Zentralisatoren sind. Archiv der Math. **6** (1954) 5—8.

GRAYEV, M.: [1] Structural isomorphisms of topological abelian groups. Rec. Math. N. S. **20** (62) (1947) 125—144.

GRECO, D.: [1] Sugli omomorfismi del reticolo dei sottogruppi normali di alcuni gruppi finiti. Ricerche Mat. **1** (1952) 241—248.

HALL, P.: [1] Complemented groups. Journal of London Math. Soc. **12** (1937) 201—204.

HIGMAN, D. G.: [1] Lattice homomorphisms induced by group homomorphisms. Proc. Amer. Math. Soc. **2** (1951) 467—478.

HONDA, K.: [1] On finite groups, whose Sylow-groups are all cyclic. Commentarii Math. Univ. Sancti Pauli **1** (1952) 5—39.

INABA, E.: [1] Über modulare Verbände, welche die Untergruppen einer endlichen abelschen Gruppe bilden. I. Proc. Imp. Acad. Japan **19** (1943) 201—204. — [2] On primary lattices. Journal of Hokkaido Univ. **11** (1948) 39—107.

ITô, N.: [1] Note on (LM)-groups of finite order. Kôdai Math. Sem. Reports (1951) 1—6.

IWASAWA, K.: [1] Über die endlichen Gruppen und die Verbände ihrer Untergruppen. Journal of Univ. of Tokyo **4-3** (1941) 171—199. — [2] On the structure of infinite M-groups. Jap. Journal of Math. **18** (1943) 709—728.

JONES, A. W.: [1] The lattice isomorphisms of certain finite groups. Duke Math. Journal **12** (1945) 541—560. — [2] Semi-modular finite groups and the BURNSIDE basis theorem. Abstract in Bull. Amer. Math. Soc. **52** (1946).

KUNTZMAN, J.: [1] Contribution à l'étude des chaînes principales d'un groupe fini. Bull. Sci. Math. **2** (71) (1947) 155—164.

KUROSH, A.: [1] Gruppentheorie. (1944), 2nd ed. (1954).

ORE, O.: [1] Structures and group theory I. Duke Math. Journal **3** (1937) 149—173. — [2] Structures and group theory II. Duke Math. Journal **4** (1938) 247—269. — [3] On the application of structure theory to groups. Bull. Amer. Math. Soc. **44** (1938) 801—806. — [4] Contributions to the theory of groups of finite order. Duke Math. Journal **5** (1939) 431—460. — [5] Remarks on structures and group relation. Vierteljschr. Naturforsch. Ges. Zürich (1940) 1—4.

PARKER, E. T.: [1] On a question raised by GARRETT BIRKHOFF. Proc. Amer. Math. Soc. **2** (1951) 901.

PERMUTTI, R.: [1] Determinazione dei gruppi finiti in omomorfism di struttura con un gruppo quadrinomio. Univ. Roma Ins. Naz. Alta. Mat. Rend. Mat. e Appl. (5) **9** (1950) 237—246. — [2] Sulle catene ad indici primi di talumi gruppi semplici. Ricerche Mat. **1** (1952) 241—248.

PETROPAVLOVSKAYA, R. V.: [1] On the determination of a group by the structure of its subsystems. Mat. Sbornik N. S. **28** (71) (1951) 63—78.

PLOTKIN, B. I.: [1] Lattice isomorphisms of soluble R-groups. Doklady Akad. SSSR. **95** (1954) 1141—1144.

RIBEIRO, H.: [1] "Lattices" des groupes abeliens finis. Comment. Math. Helv. **23** (1949) 1—17.

ROTTLÄNDER, A.: [1] Nachweis der Existenz nicht-isomorpher Gruppen von gleicher Situation der Untergruppen. Math. Zeitschrift **28** (1928) 641—653.

SADOVSKI, E. L.: [1] Über die Strukturisomorphismen von Freigruppen. Doklady **23** (1941) 171—174. — [2] Structural isomorphisms of free groups and of free products. Mat. Sbornik **14** (56) (1954) 145—173.— [3] On structural isomorphisms of free products of groups. Mat. Sbornik **21** (63) (1947) 63—82.

SATO, S.: [1] On (UM)-groups of finite order. Zenkoku-Shijô-Sûgaku-Danwakai 2–5–51 (1947). — [2] On groups and the lattices of subgroups. Osaka Math. Journal **1** (1949) 135—149. — [3] Note on lattice-isomorphisms between abelian groups and non-abelian groups. Osaka Math. Journal **3** (1951) 215—220. — [4] On the lattice homomorphisms of infinite groups. I. Osaka Math. Journal **4** (1952) 229—234. — [5] On the lattice homomorphisms of infinite groups. II. Osaka Math. Journal **6** (1954) 109—118.

SUZUKI, M.: [*1*] On the lattice of subgroups of finite groups. Trans. Amer. Math. Soc. **70** (1951) 345—371. — [*2*] On the *L*-homomorphisms of finite groups. Trans. Amer. Math. Soc. **70** (1951) 372—386.

WHITMAN, P. M.: [*1*] Lattices, equivalence relations and subgroups. Bull. Amer. Math. Soc. **52** (1946) 507—522. — [*2*] Groups with a cyclic group as lattice-homomorph. Ann. of. Math. **49** (1948) 347—351.

ZACHER, G.: [*1*] Determinazione dei gruppi finiti strutturalmente omomorfi ad un gruppo d'ordine 8 con ciclico. Rend. Semi. Mat. Univ. Padova **20** (1951) 315—328. — [*2*] Determinazione dei gruppi finiti strutturalmente omomorfi al gruppo generalizzato dei quaternioni e al gruppo abeliano d'ordine 2^ν e tipo (1, . . ., 1). ibid. **20** (1951) 346—356. — [*3*] Determinazione dei gruppi finiti strutturalamente omomorfi ad un p-gruppo hamiltoniano finito. ibid. **20** (1951) 357—364. — [*4*] Determinazione dei gruppi d'ordine finito relativamente complementari. Rend. dell'Accademia di Sc. Fis. e Mat. della Soc. Naz. di Sc. Lett. ed Arti in Napoli **19** (1952) 200—206. — [*5*] Caratterizzazione dei gruppi risolubili d'ordine finito complementari. Rend. Sem. Mat. Univ. Padova **22** (1953) 111—122.

ZAPPA, G.: [*1*] Remarks on a recent paper of O. ORE. Duke Math. Journal **6** (1940) 511—512. — [*2*] Gruppi quasi-abeliani. Pont. Acad. Sci. Acta **6** (1942) 249—267. — [*3*] Sui gruppi quasi-abeliani con elementi aperiodici. ibid. **6** (1942) 295—302. — [*4*] Caratterizzazione dei gruppi di DEDEKIND finiti. Pont. Acad. Sci. Comment. **8** (1944) 443—460. — [*5*] Sul comportamente degli elementi periodici in un gruppo di DEDEKIND infinito. Comment. Math. Helv. **18** (1945) 42—44. — [*6*] Determinazione dei gruppi finiti in omomorfismo strutturale con un gruppo ciclico. Rend. Sem. Mat. Univ. Padova **18** (1949) 140—162. — [*7*] Sulla condizione perchè un omomorfismo ordinario sia anche un omomorfismo strutturale. Giornale di Mat. Battaglini (4) **78** (1949) 182—192. — [*8*] Sulla condizione perchè un emitropismo inferiore tipico tra due gruppi sia un omotropismo. Giornale di Mat. Battaglini (4) **80** (1951) 80—101. — [*9*] Sulla risolubilità dei gruppi finiti in isomorfismo reticolare con un gruppo risolubile. Giornale di Mat. Battglini (4) **80** (1951) 213—225. — [*10*] Determinazione degli elementi neutri nel reticolo dei sottogruppi di un gruppo finito. Rend. Acad. Sci. Fis. Mat. Napoli **18** (1951) 22—28. — [*11*] Sugli omomorfismi del reticolo dei sottogruppi di un gruppo finito. Ricerche Mat. **1** (1952) 78—106.

ZASSENHAUS, H.: [*1*] Lehrbuch der Gruppentheorie. I. (1937).

ZITAROSA, A.: [*1*] Sugli elementi neutri del reticolo dei sottogruppi normali di un gruppo speciale finito. Ricerche Mat. **1** (1952) 249—254.

Index.